无人机系统应用技术系列教材

无人机应用技术开发
（DJI Payload SDK）

谢锋然　连国云　主　编

孙立晔　杨　欧　杨金锋　副主编

电子工业出版社
Publishing House of Electronics Industry
北京·BEIJING

内 容 简 介

本书遵循无人机工程师职业素养的要求和专业技能积累的规律进行内容组织，采用"项目化教学""任务驱动"方式展开讲解。本书分为 3 篇，共计 16 个项目，内容覆盖无人机与 PSDK 开发基础、PSDK 功能模块和 PSDK 综合应用。第 1 篇仅有 1 个项目，即无人机与 PSDK 开发概述；第 2 篇共有 12 个项目，即日志管理与电源管理、无人机信息管理、相机管理与云台管理、飞行控制与运动规划、自定义控件、时间同步与精准定位、SDK 互联互通、视频流传输与回放下载、健康管理系统（HMS）、获取相机码流（liveview）、本地升级、相机实现与云台实现；第 3 篇共有 3 个项目，即喊话器、空气质量检测仪、AI 目标识别。

本书是一本探讨无人机技术与应用开发的教材，适合作为无人机工程师、信息类专业的学生，以及无人机技术爱好者等人员的参考用书。

未经许可，不得以任何方式复制或抄袭本书之部分或全部内容。
版权所有，侵权必究。

图书在版编目（CIP）数据

无人机应用技术开发：DJI Payload SDK / 谢锋然，连国云主编. -- 北京：电子工业出版社，2025.4.
ISBN 978-7-121-49936-4

Ⅰ．V279

中国国家版本馆 CIP 数据核字第 2025MX9436 号

责任编辑：徐建军
印　　刷：河北鑫兆源印刷有限公司
装　　订：河北鑫兆源印刷有限公司
出版发行：电子工业出版社
　　　　　北京市海淀区万寿路 173 信箱　　邮编：100036
开　　本：787×1 092　　1/16　　印张：18.5　　字数：451 千字
版　　次：2025 年 4 月第 1 版
印　　次：2025 年 4 月第 1 次印刷
印　　数：1 200 册　　定价：59.80 元

凡所购买电子工业出版社图书有缺损问题，请向购买书店调换。若书店售缺，请与本社发行部联系，联系及邮购电话：(010) 88254888，88258888。
质量投诉请发邮件至 zlts@phei.com.cn，盗版侵权举报请发邮件至 dbqq@phei.com.cn。
本书咨询联系方式：(010) 88254570，xujj@phei.com.cn。

前言

党的二十届三中全会审议通过的《中共中央关于进一步全面深化改革、推进中国式现代化的决定》对"健全现代化基础设施建设体制机制"做出系统部署，其中专门提到"发展通用航空和低空经济"。低空经济既包括传统通用航空业态，又融合了以无人机为支撑的低空生产服务方式，是一种容纳并推动多领域协调发展的综合经济形态，具有明显的新质生产力特征。

DJI（大疆）成立于 2006 年。自成立以来，DJI（大疆）从无人机系统拓展至多元化产品体系，在无人机、手持影像系统等领域中成为全球领先品牌，以一流的技术产品重新定义"中国智造"的内涵。它在为用户带来创新、可靠的产品与解决方案的同时，迅速进入影视传媒、能源巡检、遥感测绘、农业服务和基建工程等多个领域，为各行各业提供高效、安全、智能的工具，助力行业发展，让更先进的生产、生活方式惠及更多人。同时，它致力成为公共安全和应急救援中不可或缺的中坚力量，在地震、火灾、爆炸等场景中提供强有力的支持，为保护地球生命、守护人类文明贡献力量。

为了适应时代发展的步伐，本书在众多无人机品牌中，选择 DJI（大疆）作为学习底座，在编写过程中遵循无人机工程师职业素养的要求和专业技能积累的规律，突出对职业能力、职业素养、工匠精神和质量意识的培养。从无人机研发工程师的角度出发，帮助无人机工程师开启无人机研发的"大门"。本书从实战出发讲解原理，采用真实案例构建实战场景，使用 C 语言完成代码，并结合实战进行代码解析，通过练习的方式来帮助读者加深理解。

本书由深圳职业技术大学人工智能学院组织策划，由谢锋然、连国云担任主编，由孙立晔、杨欧、杨金锋担任副主编。其中，谢锋然负责编写项目 1～项目 5，连国云负责编写项目 6～项目 9，孙立晔负责编写项目 10～项目 12，杨欧负责编写项目 13 和项目 14，杨金锋负责编写项目 15 和项目 16。书中部分项目素材由北京曾益慧创科技有限公司的刘晋东提供，本书在编写过程中参考了部分网络资源，在此一并表示感谢。

为了方便教师教学，本书配有电子教学课件及相关资源，请有此需求的教师登录华信教育资源网注册后免费下载，如有问题可在网站留言板留言或与电子工业出版社联系（E-mail：hxedu@phei.com.cn）。

希望本书能够成为您在无人机二次开发道路上的良师益友，助您拓宽视野、一探究竟、实现梦想！由于编者水平有限，书中难免存在疏漏和不足之处，希望同行专家和读者给予批评指正。

编　者

目 录

第 1 篇　无人机与 PSDK 开发基础

项目 1　无人机与 PSDK 开发概述 2
  1.1　无人机的应用与行业现状 2
    1.1.1　无人机行业现状 2
    1.1.2　无人机的组成 3
  1.2　DJI 无人机生态 3
    1.2.1　DJI SDK 介绍 3
    1.2.2　PSDK 3.x 版本 4
    1.2.3　PSDK 基础功能 4
    1.2.4　PSDK 应用场景 5
  1.3　PSDK 硬件环境 6
    1.3.1　PSDK 开发环境 6
    1.3.2　PSDK 拓展接口板 7
  1.4　PSDK 开发基础 8
    1.4.1　通用结构 8
    1.4.2　跨平台移植 10
    1.4.3　硬件接口层 11
    1.4.4　操作系统抽象层 12
    1.4.5　注册适配层 14
    1.4.6　系统初始化 17
    1.4.7　构建项目 20

第 2 篇　PSDK 功能模块

项目 2　日志管理与电源管理 28
  2.1　日志管理 28
    2.1.1　日志管理功能 28
    2.1.2　接口说明 29

2.2	电源管理		31
	2.2.1	电源管理功能	32
	2.2.2	接口说明	32
	2.2.3	树莓派 GPIO	34
2.3	代码模板		36

项目 3　无人机信息管理 48

3.1	信息管理功能	48
3.2	接口说明	50
3.3	代码模板	51

项目 4　相机管理与云台管理 60

任务 1　相机管理 60

4.1	相机管理功能	60
4.2	接口说明	61
4.3	代码模板	63

任务 2　云台管理 70

4.4	云台管理功能	70
4.5	接口说明	71
4.6	代码模板	72

项目 5　飞行控制与运动规划 79

任务 1　飞行控制 79

5.1	飞行控制功能		79
	5.1.1	飞行控制概述	79
	5.1.2	基础概念	80
5.2	接口说明		81
5.3	代码模板		82

任务 2　运动规划 89

5.4	运动规划功能		90
	5.4.1	航点任务	90
	5.4.2	工作流程	91
5.5	Waypoint v2		92
5.6	接口说明		92
5.7	代码模板		93

项目 6　自定义控件 102

6.1	App 窗口功能组件	102
6.2	自定义控件 UI	103

  6.2.1　主界面控件 103
  6.2.2　配置界面控件 103
  6.2.3　控件配置文件 104
 6.3　接口说明 107
 6.4　模板代码 109

项目 7　时间同步与精准定位 117

 任务 1　时间同步 117
  7.1　时间同步功能 117
  7.2　接口说明 118
  7.3　代码模板 119
 任务 2　精准定位 129
  7.4　精准定位功能 129
  7.5　接口说明 130
  7.6　代码模板 131

项目 8　SDK 互联互通 143

 8.1　SDK 互联互通功能 143
  8.1.1　概述 143
  8.1.2　基础概念 144
 8.2　使用 SDK 互联互通功能 145
  8.2.1　MSDK 端 145
  8.2.2　OSDK 端 146
  8.2.3　PSDK 端 149
 8.3　接口说明 151

项目 9　视频流传输与回放下载 153

 任务 1　视频流文件传输与回放下载 153
  9.1　概述 153
  9.2　数据传输 154
  9.3　配置网络参数 155
  9.4　视频流文件传输 157
  9.5　回放下载 161
  9.6　代码模板 171
 任务 2　实时视频流传输与回放下载 187
  9.7　概述 187
  9.8　H.264 编码 187
  9.9　代码模板 189

项目 10　健康管理系统（HMS） 200
10.1　健康管理系统的基本概念 200
10.2　接口说明 201
10.3　代码模板 201

项目 11　获取相机码流（liveview） 209
11.1　实时视频流 209
11.2　接口说明 210
11.3　代码模板 211

项目 12　本地升级 218
12.1　本地升级功能 218
12.2　固件版本 219
12.3　接口说明 221
12.3.1　本地升级模块 221
12.3.2　升级操作 222

项目 13　相机实现与云台实现 234
13.1　概述 234
13.2　相机功能 234
13.2.1　基本概念 235
13.2.2　注册基础功能 236
13.2.3　拍照功能 238
13.2.4　存储照片 239
13.2.5　录像功能 240
13.3　云台功能 242
13.4　代码模板 243

第 3 篇　PSDK 综合应用

项目 14　喊话器 254
14.1　音频编码基础 254
14.2　ekho（余音） 256
14.3　喊话器控件 256
14.4　使用喊话器功能 257
14.5　代码模板 262

项目 15	空气质量检测仪	269
15.1	硬件设计	269
15.2	模数转换	271

项目 16	AI 目标识别	279
16.1	实时视频流功能	279
16.2	人脸检测	279
16.3	代码模板	281

第 1 篇
无人机与 PSDK 开发基础

- 项目 1　无人机与 PSDK 开发概述

项目 1 无人机与 PSDK 开发概述

任务要求

了解 DJI 无人机的相关概念，熟悉 PSDK 的基础知识、硬件环境及开发基础，并搭建 PSDK 的开发环境。

知识导入

1.1 无人机的应用与行业现状

无人机是新一代电子信息技术与航空工业技术深度融合的产物，是全球战略性新兴科技的热门发展方向之一。作为航空产业中冉冉升起的新星，无人机产业不仅在社会生产、生活中发挥越来越重要的作用，还成为新的经济增长点。

1.1.1 无人机行业现状

无人机市场调研机构 Drone Industry Insights 在 2022 年 10 月发布了 2022 年无人机市场地图，截至发布日期，共统计到全球 1076 家与无人机相关的厂商，这些厂商提供各类硬件、软件及服务。

民用无人机产业的上游企业主要提供关键原材料和核心零部件，其中，关键原材料包括金属材料和复合材料两大类；而核心零部件一般包括芯片、电池、电机、发动机、机身结构件、陀螺仪、金属零件和复合零件等。产业中游的无人机各分系统、任务载荷和系统集成是无人机制造的核心部分。产业下游为无人机应用领域。

中国航空工业集团有限公司在2022年11月发布的《通用航空产业发展白皮书（2022）》显示，2021年全球民用无人机市场规模超过1600亿元，同比增长61.6%，其中工业级无人机占60%左右。随着下游应用领域的不断扩大，未来其市场规模将继续增长，预计2025年将达到5000亿元，届时工业级无人机市场规模占比将超过80%。民用无人机市场的发展受到技术进步、航空数据和成像需求增加，以及无人机在各行业应用范围不断扩大的影响，未来民用无人机市场很可能继续增长并扩展到新的领域。

2018年修订实施了《民用航空空中交通管理规则》。从规范性文件层面来看，近年来，民航局针对无人机运行、空中交通管理、登记、驾驶员、经营性飞行活动及适航管理等，先后出台了《轻小无人机运行规定（试行）》《民用无人机系统空中交通管理办法》《民用无人机实名制登记管理规定》《民用无人机驾驶员管理规定》《民用无人机经营性飞行活动管理办法（暂行）》《特定类无人机试运行管理规程（暂行）》《民用无人机产品适航审定管理程序（试行）》《民用无人机系统适航审定项目风险评估指南（试行）》等一系列规范性文件。

无人机的应用领域非常广泛，可以应用于军事、民用等方面。

（1）无人机可以应用于航空摄影、地质地貌测绘、森林防火、地震调查、核辐射探测、边防巡逻、抢险救灾、农作物估产、农田信息监测、高压输电线路巡查、野生动物保护、科学研究与实验、海上侦察、鱼类监测、环境监测等。

（2）无人机也可以应用于大气采样增雨、资源勘探、反恐、警察调查与巡逻、公安监测、消防航空调查、通信接力、城市规划、数字城市建设等。国内各类无人机公司主要承接无人机制造、无人机应用、无人机销售、无人机售后服务、无人机维修等业务。

1.1.2 无人机的组成

无人机（unmanned aerial vehicle）是利用无线遥控或程序控制来执行特定航空任务的飞行器，是指不搭载操作人员的一种动力空中飞行器，采用空气动力为飞行器提供所需的升力，能够自动飞行或远程引导；既能一次性使用又能进行回收，还能携带致命性和非致命性有效负载。

无人机按照用途可分为无人侦察机、电子战无人机、靶机、反辐射无人机、对地攻击无人机、通信中继无人机、火炮较射无人机、特种无人机、诱饵无人机等；按照飞行方式可分为固定翼无人机、旋翼无人机、扑翼无人机、飞艇等。

DJI无人机系统一般包括地面系统、飞机系统、任务载荷和无人机使用保障人员。

1.2 DJI无人机生态

1.2.1 DJI SDK 介绍

DJI开发者平台提供了多种SDK，开发者可以根据需求开发自己的无人机解决方案。

1. Payload SDK

DJI 为支持开发者开发出可挂载在 DJI 无人机上的负载设备，提供了开发工具包 Payload SDK（PSDK）。它允许开发者自行开发满足自身需求的各种负载设备，并提供了一整套接口，以实现负载设备与无人机之间的各种通信，这需要开发者具备硬件方面的知识。

2. Onboard SDK

Onboard SDK（OSDK）是一个用于开发无人机应用程序的开发工具包，基于 OSDK 开发的应用程序能够运行在机载计算机上（如 Manifold 2）。开发者通过调用 OSDK 中的指定接口能够获取无人机上的各类数据，经过开发者设计的软件逻辑和算法框架，可进行相应的计算和处理，生成对应的控制指令来控制无人机执行相应的动作，实现自动化飞行、负载控制和视频图像分析等功能。

3. Mobile SDK

Mobile SDK（MSDK）是一款软件开发套件，旨在让开发者能够访问 DJI 无人机和手持相机产品的丰富功能。MSDK 通过兼顾底层的功能，如飞行稳定、电池管理、信号传输和通信等，简化了应用程序的开发过程。这样，开发者就不需要具备丰富的机器人或嵌入式系统的背景知识，专注与 DJI 产品相关的行业应用开发即可。开发者可以通过 Mobile SDK 访问许多 DJI 产品的功能，也可以实现自主飞行、控制相机和云台、接收实时视频图传和传感器数据、下载保存好的媒体文件，以及监听其他组件的状态。

1.2.2　PSDK 3.x 版本

PSDK 3.x 版本集 OSDK、PSDK 功能于一体，结构更加清晰易懂。对于新接触 PSDK 功能的开发者，推荐使用 PSDK 3.x 之后的新版本进行开发，这也是官方的主要维护版本，新手初学及开发之后的维护也会更容易。

1.2.3　PSDK 基础功能

PSDK 开发的设备可以作为符合 DJI 无人机接入标准的负载。对飞机本体而言，挂载在无人机上的设备均可以作为负载，如 DJI 官方相机就是无人机的负载相机设备。负载设备的功能极大地拓展了无人机应用，对于不同的开发者或开发商企业，其开发的设备的标准或型号不尽相同。将它们安装到无人机上会缺乏统一的标准，可能影响飞行安全，而且不能直接使用 DJI 遥控器或 App 对负载进行可视化操作。PSDK 可以被认为是负载设备接入 DJI 无人机的标准，包括硬件标准和软件标准。PSDK 将规定负载接入标准，通过 SDK API 的方式将负载功能抽象出来，DJI 官方的 App 或自行开发的 MSDK 就可以直接进行操控或数据交互，并进行功能拓展。

PSDK 基础功能模块如表 1-1 所示。

表 1-1　PSDK 基础功能模块

功能模块	功能描述
日志管理	拥有日志管理功能的负载设备支持用户通过串口、终端或 USB 等输出不同模块的日志信息
信息管理	用于无人机系统的信息获取和消息订阅
相机功能	设置相机模式、拍照、录像、获取相机状态、指点变焦、测光、对焦、视频流传输、媒体文件回放与下载
相机管理	设置相机的光圈、曝光时间及分辨率等各项参数；控制相机实现拍照、录像及指点变焦等功能；支持单独或批量下载、删除相机的原始照片和视频数据
云台功能	控制云台转动速度和转动方式；设置平滑系数和最大速度系数；支持云台校准、复位及角度微调
云台管理	控制云台转动的角度和角速度；获取云台当前的角度和角速度
电源管理	具有电源管理功能的负载设备不仅能够向无人机申请较高的功率，还能够接收无人机发送的关机通知并正常关机，防止负载设备意外丢失数据信息
飞行控制	设置或获取无人机飞行控制器的参数；执行基本的飞行任务；设置无人机基础控制功能
自定义控件	用于负载控制、浮窗信息显示
HMS 功能	获取无人机的健康管理信息，有效监控无人机的当前状态
时间同步	获取无人机的硬件触发脉冲信号；获取统一的 UTC 时间
数据传输	用于数据传输、状态获取、带宽控制
X-PORT 控制	获取 X-PORT 状态；控制云台转动速度和转动方式；设置 X-PORT 自定义限位角
高级视觉	获取相机的 RGB 码流和 H.264 码流，用户可对视频流解码进行实时处理
本地升级	升级负载设备固件

1.2.4　PSDK 应用场景

DJI 为支持开发者开发出可挂载在 DJI 无人机上的负载设备，提供了开发工具包 PSDK，以及开发配件 X-PORT 标准云台、SkyPort v2 转接环和 SDK 同轴线等，方便开发者利用 DJI 无人机上的电源、通信链路及状态信息等资源。开发者能够根据行业的应用需求，基于 PSDK 提供的功能接口，并结合具体的结构设计、硬件设计、软件逻辑实现和算法优化，开发出自动巡检系统、红外相机、测绘相机、多光谱相机、喊话器、探照灯等满足不同细分领域的负载设备。

使用 PSDK 开发可挂载在 DJI 无人机上的负载设备，能够满足不同行业多样化的应用需求。PSDK 负载应用场景如表 1-2 所示。

表 1-2　PSDK 负载应用场景

负载设备	安防			巡检		勘测			环保		更多
	治安管理	消防	应急救援	管网巡检	厂区巡检	地质测绘	空间规划	资源勘探	生态监测	生物保护	……
可变焦相机	✓	✓	✓	✓	✓	✓	✓	✓	✓	✓	…
热成像相机	✓	✓	✓	✓	✓			✓	✓	✓	…
红外相机	✓	✓	✓		✓			✓	✓	✓	…
多目相机	✓	✓	✓			✓	✓	✓			…

续表

负载设备	安防			巡检		勘测			环保		更多
	治安管理	消防	应急救援	管网巡检	厂区巡检	地质测绘	空间规划	资源勘探	生态监测	生物保护	……
星光相机	✓		✓								…
激光雷达			✓	✓		✓	✓				…
气体检测仪			✓	✓	✓				✓	✓	…
辐射检测仪	✓	✓					✓	✓			…
水质检测仪									✓		…
喊话器	✓	✓	✓	✓	✓						…
探照灯	✓	✓	✓		✓			✓			…
自动巡检系统				✓	✓						…
更多负载	…	…	…	…	…	…	…	…	…	…	…

1.3 PSDK 硬件环境

1.3.1 PSDK 开发环境

DJI 无人机具有一系列的机型，不同机型在功能上有一定的差异，开发者需要根据操作系统和开发平台对 PSDK 功能的支持差异、负载设备程序的资源占用情况及 PSDK 支持的工具链，选择开发负载设备的操作系统和开发平台。

1. 选购无人机

有关无人机的参数和使用无人机的详细说明参见官网中所选购机型的快速入门指南。

2. 开发套件

1）X-PORT

开发者借助 X-PORT 能够快速开发出可挂载在 DJI 无人机上且具有云台功能的负载设备。

2）SkyPort v2

开发者借助 SkyPort v2 开发套件拓展接口板能够使用第三方开发板开发负载设备。

3）第三方开发平台

DJI PSDK 使用 STM3240G-EVAL（STM32F407IG）开发板开发并调试示例程序。建议开发者参考 STM3240G-EVAL（STM32F407IG）的参数信息选购所需的开发平台。

3. 操作系统

PSDK 开发支持 Linux 操作系统和 RTOS 操作系统。对于不同操作系统的功能差异，请查阅大疆官网。在使用 Linux 操作系统开发环境时，需要安装 C 编译器（GCC 5.4.0/5.5.0 版本）、CMake（2.8 及以上版本）、FFmpeg（版本需要高于或等于 4.1.3 且低于 5.0.0）等开发工具。

4. 开发平台

PSDK 支持使用一系列工具编译基于 PSDK 开发的负载设备，请根据选用的开发平台正确选择工具链。开发者需要根据所选用的开发平台，选择指定编译链的静态库。PSDK 开发平台如图 1-1 所示。

工具链名称	目标平台	典型芯片型号	推荐开发平台
aarch64-linux-gnu-gcc	aarch64-linux-gnu	NVIDIA Jetson TX2、Rockchip RK3399 pro	Manifold2-G、瑞芯微 Toybrick开发板
x86_64-linux-gnu-gcc	x86_64-linux-gnu	64位intel处理器，如 Intel Core i7-8550U	Maniflod2-C
arm-linux-gnueabi-gcc	arm-linux-gnueabi	ZYNQ、I.MX6Q	-
arm-linux-gnueabihf-gcc	arm-linux-gnueabihf	支持硬件浮点运算的处理器，如OK5718-C等	-
armcc-cortex-m4	Cortex M4/M4F系列 MCU	STM32F407IGT6、STM32F405RGT6	STM32F407-Eval、STM32F407探索者开发板等
arm-none-eabi-gcc	Cortex M4/M4F系列 MCU	STM32F407IGT6、STM32F405RGT6	STM32F407-Eval、STM32F407探索者开发板等
arm-linux-androideabi-gcc	arm-linux-androideabi	高通骁龙系列芯片	安卓平台
aarch64-linux-android-gcc	aarch64-linux-android	高通骁龙系列芯片	安卓平台

图 1-1　PSDK 开发平台

1.3.2　PSDK 拓展接口板

1. 硬件环境

PSDK 在无人机上的接口是 DJI 标准云台接口。M300 上的负载云台接口是连接到 Type-C 接口的，但是 DJI PSDK 的 Type-C 接口与 OSDK 的不一样，引脚线序定义不对开发者开放，所以 DJI 负载设备必须连接到官方的云台接口。

SkyPort 或 X-PORT 是接入 DJI 云台接口的标准接口，是 PSDK 负载必需的配件。PSDK 负载需要有自己的 CPU/MCU，且可以支持 PSDK lib。PSDK 3.0 中 PSDK 的 lib 支持与 OSDK 的 lib 支持是一致的，PSDK 与无人机通信是基于串口和网口（Linux）的，PSDK 负载平台还需要支持串口和网络 socket 通信。不支持网络的平台，将不支持部分用到网络 socket 通信的功能，如 MOP、高速数据传输。

综上所述，PSDK 开发阶段需要的配件如下。
- 无人机负载云台支架（无人机自带）。
- SkyPort 或 X-PORT。
- 支持 PSDK lib 的机载平台，且支持串口和网络 socket 通信。

2. SkyPort

SkyPort 是用于连接 DJI 云台接口的标准转接环，分为 SkyPort v1 和 SkyPort v2。SkyPort v2 提供了排线接口和同轴线接口，使用排线通过 Port 1 接口可将 SkyPort v2 连接至拓展接口板或第三方开发板；使用同轴线通过 Port 2 接口可将 SkyPort v2 连接至 SkyPort 负载设备开发板。

3. 拓展接口板

SkyPort/X-PORT 连接负载的接口是引脚定义的，为了方便开发，需要将引脚拓展成标准的接口。官方提供的拓展接口板用于将排线引脚或同轴线引脚拓展成串口 TTL 引脚、RJ45 网口，并通过转压将无人机提供的电源转压。同时，拓展接口板提供 9V/2A、5V/2A，以及与引脚自身相关的电压 13.6V/17V 的接口。

在使用 SkyPort v2 时，可以用同轴线或排线，且只能用一根连接线。在使用 X-PORT 时，用排线连接拓展接口板即可。拓展接口板如图 1-2 所示。

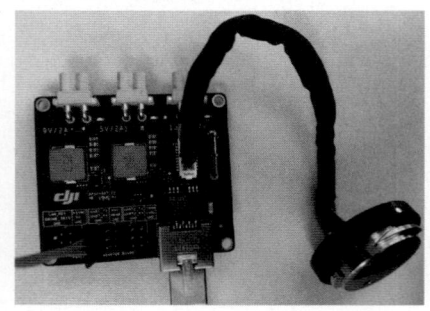

图 1-2　拓展接口板

1.4　PSDK 开发基础

1.4.1　通用结构

为了方便开发者使用与 PSDK 开发相关的应用，我们将 PSDK 的通用部分独立出来，在使用时直接调用即可，通用结构如下。

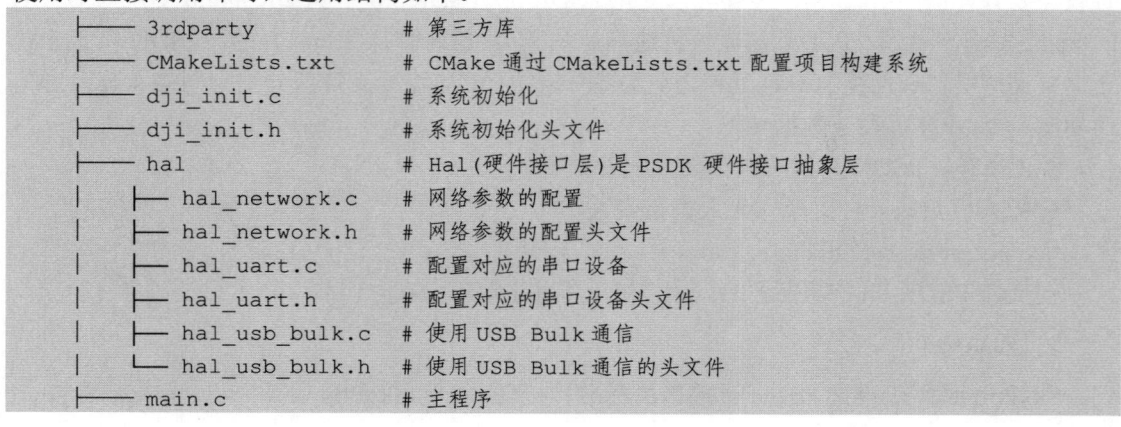

```
├── monitor                              # 监控
│   ├── sys_monitor.c
│   └── sys_monitor.h
├── osal                                 # Osal(操作系统抽象层)是 PSDK 的操作系统抽象层
│   ├── osal.c                           # 使用标准库 pthread 封装 T_DjiOsalHandler 中的线程函数、互
斥锁、信号量及时间接口等
│   ├── osal_fs.c
│   ├── osal_fs.h
│   ├── osal.h
│   ├── osal_socket.c
│   └── osal_socket.h
└── psdk_lib                             # psdk 库文件
    ├── include
    │   ├── dji_aircraft_info.h          # 无人机信息(信息管理)相关功能的头文件
    │   ├── dji_camera_manager.h         # PSDK 相机相关功能的头文件
    │   ├── dji_core.h                   # PSDK Core 相关功能的头文件
    │   ├── dji_error.h                  # 错误码的头文件
    │   ├── dji_fc_subscription.h        # 消息订阅相关功能的头文件
    │   ├── dji_flight_controller.h      # 飞行控制相关功能的头文件
    │   ├── dji_gimbal.h                 # PSDK 云台相关功能的头文件
    │   ├── dji_gimbal_manager.h         # PSDK 云台管理相关功能的头文件
    │   ├── dji_high_speed_data_channel.h # 高速数据通道相关功能的头文件
    │   ├── dji_hms.h                    # 健康管理系统相关功能的头文件
    │   ├── dji_hms_info_table.h         # HMS 信息表格功能的头文件
    │   ├── dji_liveview.h               # 获取相机码流相关功能的头文件
    │   ├── dji_logger.h                 # 日志管理相关功能的头文件
    │   ├── dji_low_speed_data_channel.h # 低速数据通道相关功能的头文件
    │   ├── dji_mop_channel.h            # SDK 互联互通相关功能的头文件
    │   ├── dji_payload_camera.h         # PSDK 相机相关功能的头文件
    │   ├── dji_perception.h             # 获取感知灰度图相关功能的头文件
    │   ├── dji_platform.h               # 跨平台相关功能的头文件
    │   ├── dji_positioning.h            # 精准定位相关功能的头文件
    │   ├── dji_power_management.h       # PSDK 电源管理相关功能的头文件
    │   ├── dji_time_sync.h              # 时间同步相关功能的头文件
    │   ├── dji_typedef.h                # 宏定义和结构体的关键信息与使用方法的头文件
    │   ├── dji_upgrade.h                # 本地升级相关功能的头文件
    │   ├── dji_version.h                # PSDK 获取版本信息相关功能的头文件
    │   ├── dji_Waypoint_v2.h            # Waypoint v2 相关功能的头文件
    │   ├── dji_Waypoint_v2_type.h       # Waypoint v2 type 相关功能的头文件
    │   ├── dji_Waypoint_v3.h            # Waypoint v3 相关功能的头文件
    │   ├── dji_Widget.h                 # 自定义控件相关功能的头文件
    │   ├── dji_xport.h                  # X-PORT 控制相关功能的头文件
    │   └── legacy_psdk2.x               # 此目录包含了 PSDK 2.x 版本的头文件,省略
    └── lib
        └── arm-linux-gnueabihf-gcc      # 满足树莓派
            └── libpayloadsdk.a          # psdk 静态链接库
```

psdk_lib（库与头文件）的 include 文件夹中包含各功能模块的头文件，而 lib 文件夹中包含不同平台的静态库。PSDK 3.x 版本支持的硬件平台如表 1-3 所示。

表 1-3　PSDK 3.x 版本支持的硬件平台

硬件平台	链接库名
aarch64-himix100-linux-gcc	libpayloadsdk.a
arm-himix100-linux-gcc	libpayloadsdk.a
arm-hisiv500-linux-gcc	libpayloadsdk.a
arm-linux-gnueabihf-gcc	libpayloadsdk.a
aarch64-linux-android-gcc	libpayloadsdk.a
arm-himix200-linux-gcc	libpayloadsdk.a
arm-hisiv600-linux-gcc	libpayloadsdk.a
arm-none-eabi-gcc	libpayloadsdk.a
aarch64-linux-gnu-gcc	libpayloadsdk.a
arm-hisiv300-linux-gcc	libpayloadsdk.a
arm-linux-androideabi-gcc	libpayloadsdk.a
x86_64-linux-gnu-gcc	libpayloadsdk.a
armcc_cortex-m4	libpayloadsdk.a
arm-hisiv400-linux-gcc	libpayloadsdk.a
arm-linux-gnueabi-gcc	libpayloadsdk.a
xtensa-esp32-elf-gcc	libpayloadsdk.a

本书使用树莓派开发板，执行以下命令输出树莓派开发板的硬件架构名称是 armv7l。PSDK 硬件平台选择了 arm-linux-gnueabihf-gcc。

```
pi@raspberrypi:~/jz/pro_3_fc/build $ uname -m
armv7l
```

1.4.2　跨平台移植

为了基于 PSDK 开发的负载设备控制程序能移植到不同版本的软硬件平台中，需要通过 Hal（hardware abstraction layer，硬件接口层）适配不同的硬件平台，并通过 Osal（operating system abstraction layer，操作系统抽象层）实现与不同操作系统的兼容。跨平台移植如图 1-3 所示。

图 1-3　跨平台移植

在将基于 PSDK 开发的负载设备控制程序移植到不同版本的软硬件平台上时，首先，需要初始化 Hal 层和 Osal 层，并注册关键的配置信息；然后，通过加载静态库、引用指定的资源文件并声明结构体，设置负载设备控制程序跨平台移植所需的配置信息；最后，使用指定的接口将 Platform 模块注册到负载设备控制程序中，获取硬件资源和操作系统资源，实现负载设备控制程序的跨平台移植。

1.4.3 硬件接口层

硬件接口层是 PSDK 硬件接口抽象层，位于负载设备控制程序和硬件资源之间。

开发者需要按照 DjiPlatform_RegHalUartHandler、DjiPlatform_RegHalNetworkHandler 与 DjiPlatform_RegHalUsbBulkHandler 接口中的函数原型，实现并将适配 Hal 层的函数注册到负载设备控制程序中，使基于 PSDK 开发的负载设备控制程序通过 Hal 层即可直接访问负载设备硬件的底层资源，控制负载设备执行相应的动作，使负载设备控制程序能够适配不同的硬件平台，如 STM32F407IGH6-EVAL 或 Manifold 2-C 等。

1. 网口设备

使用网口的设备适配 Hal 层函数需要执行如下操作。

（1）实现适配 Hal 层网口操作函数。

- 本地网卡初始化：

```
T_DjiReturnCode (*NetworkInit)(const char *ipAddr,
const char *netMask,
T_DjiNetworkHandle *networkHandle)
```

- 本地网卡反初始化：

```
T_DjiReturnCode (*NetworkDeInit)(T_DjiNetworkHandle networkHandle)
```

- 获取网卡信息：

```
T_DjiReturnCode (*NetworkGetDeviceInfo)(
T_DjiHalNetworkDeviceInfo *deviceInfo)
```

（2）使用 DjiPlatform_RegHalUartHandler 接口注册网口操作函数。

2. 串口设备

使用串口通信的设备适配 Hal 层函数需要执行如下操作。

（1）实现适配 Hal 层 UART 操作函数。

- 串口初始化：

```
T_DjiReturnCode (*UartInit)(E_DjiHalUartNum uartNum,
uint32_t baudRate,
T_DjiUartHandle *uartHandle)
```

- 串口反初始化：

```
T_DjiReturnCode (*UartDeInit)(T_DjiUartHandle uartHandle)
```

- 发送数据：

```
T_DjiReturnCode (*UartWriteData)(T_DjiUartHandle uartHandle,
const uint8_t *buf,
uint32_t len,
uint32_t *realLen)
```

- 接收数据：

```
T_DjiReturnCode (*UartReadData)(T_DjiUartHandle uartHandle,
uint8_t *buf,
uint32_t len,
uint32_t *realLen)
```

- 获取串口状态：

```
T_DjiReturnCode (*UartGetStatus)(E_DjiHalUartNum uartNum, T_DjiUartStatus *status)
```

（2）使用 DjiPlatform_RegHalNetworkHandler 接口注册 UART 操作函数。

说明：负载设备串口的参数如下。
- 波特率：460800。
- 停止位：1。
- 数据位：8。
- 奇偶校验：无。

3. USB 设备

使用 USB Bulk 通信的设备适配 Hal 层函数需要执行如下操作。

（1）实现适配 Hal 层 USB Bulk 操作函数。

- USB Bulk 初始化：

```
T_DjiReturnCode (*UsbBulkInit)(T_DjiHalUsbBulkInfo usbBulkInfo, T_DjiUsbBulkHandle *usbBulkHandle)
```

- USB Bulk 反初始化：

```
T_DjiReturnCode (*UsbBulkDeInit)(T_DjiUsbBulkHandle usbBulkHandle)
```

- USB Bulk 写数据：

```
T_DjiReturnCode (*UsbBulkWriteData)(T_DjiUsbBulkHandle usbBulkHandle, const uint8_t *buf, uint32_t len, uint32_t *realLen)
```

- USB Bulk 读数据：

```
T_DjiReturnCode (*UsbBulkReadData)(T_DjiUsbBulkHandle usbBulkHandle, uint8_t *buf, uint32_t len, uint32_t *realLen)
```

- 获取 USB Bulk 信息：

```
T_DjiReturnCode (*UsbBulkGetDeviceInfo)(T_DjiHalUsbBulkDeviceInfo *deviceInfo)
```

（2）使用 DjiPlatform_RegHalUsbBulkHandler 接口注册 USB Bulk 操作函数。

1.4.4 操作系统抽象层

操作系统抽象层是 PSDK 的操作系统抽象层，位于负载设备控制程序和操作系统之间。开发者需要按照 DjiPlatform_RegOsalHandler 接口中的函数原型，实现并将适配不同操作系统的函数注册到负载设备控制程序中。使用 PSDK 开发的负载设备控制程序可直接访问操作系统及操作系统内核的资源，从而便于将负载设备控制程序移植到不同的操作系统上。

1. 线程函数

当开发者使用线程机制管理负载设备控制程序执行相应的任务时，需要实现创建线程、销毁线程和线程睡眠的函数。

- 创建线程函数：

```
T_DjiReturnCode (*TaskCreate)(const char *name,
                              void *(*taskFunc)(void *),
```

```
                        uint32_t stackSize,
                        void *arg,
                        T_DjiTaskHandle *task)
```
- 销毁线程函数：
```
T_DjiReturnCode (*TaskDestroy)(T_DjiTaskHandle task)
```
- 线程睡眠函数：
```
T_DjiReturnCode (*TaskSleepMs)(uint32_t timeMs)
```

2. 互斥锁

互斥锁是一种用于防止多个线程同时对同一队列、计数器或中断处理程序等公共资源执行读写操作的机制，能够有效避免进程死锁或长时间的等待。当开发者使用互斥锁机制时，需要实现创建互斥锁、销毁互斥锁、互斥锁上锁和互斥锁解锁的函数。

- 创建互斥锁函数：
```
T_DjiReturnCode (*MutexCreate)(T_DjiMutexHandle *mutex)
```
- 销毁互斥锁函数：
```
T_DjiReturnCode (*MutexDestroy)(T_DjiMutexHandle mutex)
```
- 互斥锁上锁函数：
```
T_DjiReturnCode (*MutexLock)(T_DjiMutexHandle mutex)
```
- 互斥锁解锁函数：
```
T_DjiReturnCode (*MutexUnlock)(T_DjiMutexHandle mutex)
```

3. 信号量

信号量是一种用于防止多线程同时操作相同代码段的机制。当开发者使用信号量机制时，需要实现创建信号量、销毁信号量、等待信号量、等待超时信号量和释放信号量的函数。

- 创建信号量函数：
```
T_DjiReturnCode (*SemaphoreCreate)(uint32_t initValue, T_DjiSemaHandle *semaphore)
```
说明：在使用该函数时，请设置 initValue 信号量的初始值。

- 销毁信号量函数：
```
T_DjiReturnCode (*SemaphoreDestroy)(T_DjiSemaHandle semaphore)
```
- 等待信号量函数：
```
T_DjiReturnCode (*SemaphoreWait)(T_DjiSemaHandle semaphore)
```
说明：等待信号量函数等待时间的最大值为 32767 ms。

- 等待超时信号量函数：
```
T_DjiReturnCode (*SemaphoreTimedWait)(T_DjiSemaHandle semaphore, uint32_t waitTimeMs)
```
- 释放信号量函数：
```
T_DjiReturnCode (*SemaphorePost)(T_DjiSemaHandle semaphore)
```

4. 时间接口

- 获取当前系统的时间（ms）：
```
T_DjiReturnCode (*GetTimeMs)(uint32_t *ms)
```

- 获取当前系统的时间（us）：

```
T_DjiReturnCode (*GetTimeUs)(uint64_t *us)
```

5. 内存管理接口

- 申请内存：

```
void *(*Malloc)(uint32_t size)
```

- 释放内存：

```
void (*Free)(void *ptr)
```

1.4.5 注册适配层

跨平台移植模块必须在其他 PSDK 功能模块前进行注册，如果注册失败或未注册，则开发者将无法使用基于 PSDK 开发的负载设备。在注册前，需要设置 hal 和 osal 的相关参数。

1. 设置 hal 参数

- 设置 hal 串口参数（T_DjiHalUartHandler）。
- 设置 hal 网络参数（T_DjiHalNetworkHandler）。
- 设置 hal_usb 集成参数（T_DjiHalUsbBulkHandler）。

```
/*--------设置 hal 串口参数--------*/
T_DjiHalUartHandler uartHandler = {
    .UartInit = HalUart_Init,
    .UartDeInit = HalUart_DeInit,
    .UartWriteData = HalUart_WriteData,
    .UartReadData = HalUart_ReadData,
    .UartGetStatus = HalUart_GetStatus,
};

/*--------设置 hal 网络参数--------*/
T_DjiHalNetworkHandler networkHandler = {
    .NetworkInit = HalNetWork_Init,
    .NetworkDeInit = HalNetWork_DeInit,
    .NetworkGetDeviceInfo = HalNetWork_GetDeviceInfo,
};

/*--------设置 hal_usb 集成参数--------*/
T_DjiHalUsbBulkHandler usbBulkHandler = {
    .UsbBulkInit = HalUsbBulk_Init,
    .UsbBulkDeInit = HalUsbBulk_DeInit,
    .UsbBulkWriteData = HalUsbBulk_WriteData,
    .UsbBulkReadData = HalUsbBulk_ReadData,
    .UsbBulkGetDeviceInfo = HalUsbBulk_GetDeviceInfo,
};
```

2. 设置 osal 参数

- 设置 osal 平台参数（T_DjiOsalHandler）。
- 设置 osal 系统参数（T_DjiFileSystemHandler）。
- 设置 osal 网络参数（T_DjiSocketHandler）。

```c
/*--------设置osal平台参数--------*/
    T_DjiOsalHandler osalHandler = {
        .TaskCreate = Osal_TaskCreate,
        .TaskDestroy = Osal_TaskDestroy,
        .TaskSleepMs = Osal_TaskSleepMs,
        .MutexCreate= Osal_MutexCreate,
        .MutexDestroy = Osal_MutexDestroy,
        .MutexLock = Osal_MutexLock,
        .MutexUnlock = Osal_MutexUnlock,
        .SemaphoreCreate = Osal_SemaphoreCreate,
        .SemaphoreDestroy = Osal_SemaphoreDestroy,
        .SemaphoreWait = Osal_SemaphoreWait,
        .SemaphoreTimedWait = Osal_SemaphoreTimedWait,
        .SemaphorePost = Osal_SemaphorePost,
        .Malloc = Osal_Malloc,
        .Free = Osal_Free,
        .GetTimeMs = Osal_GetTimeMs,
        .GetTimeUs = Osal_GetTimeUs,
};

    /*--------设置osal系统参数--------*/
    T_DjiFileSystemHandler fileSystemHandler = {
        .FileOpen = Osal_FileOpen,
        .FileClose = Osal_FileClose,
        .FileWrite = Osal_FileWrite,
        .FileRead = Osal_FileRead,
        .FileSync = Osal_FileSync,
        .FileSeek = Osal_FileSeek,
        .DirOpen = Osal_DirOpen,
        .DirClose = Osal_DirClose,
        .DirRead = Osal_DirRead,
        .Mkdir = Osal_Mkdir,
        .Unlink = Osal_Unlink,
        .Rename = Osal_Rename,
        .Stat = Osal_Stat,
    };

    /*--------设置osal网络参数--------*/
    T_DjiSocketHandler socketHandler = {
        .Socket = Osal_Socket,
        .Bind = Osal_Bind,
```

```
        .Close = Osal_Close,
        .UdpSendData = Osal_UdpSendData,
        .UdpRecvData = Osal_UdpRecvData,
        .TcpListen = Osal_TcpListen,
        .TcpAccept = Osal_TcpAccept,
        .TcpConnect = Osal_TcpConnect,
        .TcpSendData = Osal_TcpSendData,
        .TcpRecvData = Osal_TcpRecvData,
    };
```

3. 注册配置信息

首先，要确保上述 hal 和 osal 的参数配置正确，否则所注册的接口不能正常使用；然后，注册配置信息。

```
    /*--------注册 Linux 操作系统中间件--------*/
    returnCode = DjiPlatform_RegOsalHandler(&osalHandler);
    if (returnCode != DJI_ERROR_SYSTEM_MODULE_CODE_SUCCESS) {
        printf("register osal handler error");
        return DJI_ERROR_SYSTEM_MODULE_CODE_SYSTEM_ERROR;
    }

    /*--------注册串口硬件--------*/
    returnCode = DjiPlatform_RegHalUartHandler(&uartHandler);
    if (returnCode != DJI_ERROR_SYSTEM_MODULE_CODE_SUCCESS) {
        printf("register hal uart handler error");
        return DJI_ERROR_SYSTEM_MODULE_CODE_SYSTEM_ERROR;
    }

    /*--------注册网络硬件--------*/
    returnCode = DjiPlatform_RegHalNetworkHandler(&networkHandler);
    if (returnCode != DJI_ERROR_SYSTEM_MODULE_CODE_SUCCESS) {
        printf("register hal network handler error");
        return DJI_ERROR_SYSTEM_MODULE_CODE_SYSTEM_ERROR;
    }

    /*--------注册 osal 网络--------*/
    returnCode = DjiPlatform_RegSocketHandler(&socketHandler);
    if (returnCode != DJI_ERROR_SYSTEM_MODULE_CODE_SUCCESS) {
        printf("register osal socket handler error");
        return DJI_ERROR_SYSTEM_MODULE_CODE_SYSTEM_ERROR;
    }

    /*--------注册 osal 系统参数--------*/
    returnCode = DjiPlatform_RegFileSystemHandler(&fileSystemHandler);
    if (returnCode != DJI_ERROR_SYSTEM_MODULE_CODE_SUCCESS) {
        printf("register osal filesystem handler error");
        return DJI_ERROR_SYSTEM_MODULE_CODE_SYSTEM_ERROR;
    }
```

1.4.6 系统初始化

dji_init.c 用于系统初始化,将完成系统环境的创建,包括配置 hal 参数与 osal 参数、注册与加载用户信息、初始化 DJI PSDK,以及定义通信、日志文件等相关参数。

1. 加入头文件

```
#include "dji_init.h"
#include <dji_typedef.h>
#include <dji_core.h>
#include <dji_aircraft_info.h>
#include <dji_logger.h>
#include <errno.h>
#include <utils/util_misc.h>
#include "osal/osal.h"
#include "osal/osal_fs.h"
#include "osal/osal_socket.h"
#include "hal/hal_uart.h"
#include "hal/hal_network.h"
#include "hal/hal_usb_bulk.h"
```

2. 设置参数

```
/*--------定义通信参数--------*/
#define DJI_USE_ONLY_UART                   (0)
#define DJI_USE_UART_AND_USB_BULK_DEVICE    (1)
#define DJI_USE_UART_AND_NETWORK_DEVICE     (2)
#define CONFIG_HARDWARE_CONNECTION          DJI_USE_UART_AND_NETWORK_DEVICE

/*--------定义日志参数--------*/
#define DJI_LOG_PATH                    "Logs/DJI"
#define DJI_LOG_INDEX_FILE_NAME         "Logs/latest"
#define DJI_LOG_FOLDER_NAME             "Logs"
#define DJI_LOG_PATH_MAX_SIZE           (128)
#define DJI_LOG_FOLDER_NAME_MAX_SIZE    (32)
#define DJI_LOG_MAX_COUNT               (10)
#define DJI_SYSTEM_CMD_STR_MAX_SIZE     (64)

/*--------定义用户参数,用户根据本人的账号填写--------*/
#define USER_App_NAME           "RaspberryPi"
#define USER_App_ID             "126413"
#define USER_App_KEY            "0ea2eadb314d1a6e............"
#define USER_App_LICENSE        "HZ/GPtycW="
#define USER_DEVELOPER_ACCOUNT  "abc@outlo**.com"
#define USER_BAUD_RATE          "460800"

/*--------创建文件对象--------*/
```

```
static FILE *s_djiLogFile;              // 创建log文件对象
static FILE *s_djiLogFileCnt;           // 创建error日志文件对象
```

3. 定义私有函数

这里共定义了5个私有函数。

```
/*------功能：设置和注册hal参数与osal参数--------*/
static T_DjiReturnCode DjiUser_PrepareSystemEnvironment(void);

/*------功能：注册用户参数--------*/
static T_DjiReturnCode DjiUser_FillInUserInfo(T_DjiUserInfo *userInfo);

/*------功能：控制台输出信息--------*/
static T_DjiReturnCode DjiUser_PrintConsole(const uint8_t *data, uint16_t dataLen);

/*------功能：写入本地文件--------*/
static T_DjiReturnCode DjiUser_LocalWrite(const uint8_t *data, uint16_t dataLen);
static T_DjiReturnCode DjiUser_LocalWriteFsInit(const char *path);
```

4. 初始化函数

初始化函数system_init将完成整个系统的初始化。

```
int system_init()
{
    T_DjiReturnCode returnCode;
    T_DjiUserInfo userInfo;
    T_DjiAircraftInfoBaseInfo aircraftInfoBaseInfo;

    /*--------创建系统环境--------*/
    returnCode = DjiUser_PrepareSystemEnvironment();
    if (returnCode != DJI_ERROR_SYSTEM_MODULE_CODE_SUCCESS) {
        USER_LOG_ERROR("Prepare system environment error");
        return DJI_ERROR_SYSTEM_MODULE_CODE_SYSTEM_ERROR;
    }

    /*--------加载用户信息--------*/
    returnCode = DjiUser_FillInUserInfo(&userInfo);
    if (returnCode != DJI_ERROR_SYSTEM_MODULE_CODE_SUCCESS) {
        USER_LOG_ERROR("Fill user info error, please check user info config");
        return DJI_ERROR_SYSTEM_MODULE_CODE_SYSTEM_ERROR;
    }

    /*--------初始化DJI PSDK--------*/
    returnCode = DjiCore_Init(&userInfo);
    if (returnCode != DJI_ERROR_SYSTEM_MODULE_CODE_SUCCESS) {
        USER_LOG_ERROR("Core init error");
        return DJI_ERROR_SYSTEM_MODULE_CODE_SYSTEM_ERROR;
```

        }
        return 0;
}
```

### 5. 主函数模板

main 函数是整个项目的主函数,开发者可以根据需要实现的功能添加相应代码。

```c
/*--------加入通用的头文件--------------------*/
#include <unistd.h>
#include <dji_core.h>
#include <dji_logger.h>
#include <utils/util_misc.h>
#include "dji_init.h"

/*---开始:在此处加入需要实现功能的头文件,下面以消息订阅的头文件为例----*/
#include <fc_subscription/fc_subscription.h>

/*---结束:在此处加入需要实现功能的头文件,下面以消息订阅的头文件为例----*

/*-----------main 函数为主函数入口---------------------*/
int main(int argc, char **argv)
{
    T_DjiReturnCode returnCode;

    USER_UTIL_UNUSED(argc);
    USER_UTIL_UNUSED(argv);

    /*-----------系统初始化-----------------------*/
    returnCode = system_init();
    if (returnCode != DJI_ERROR_SYSTEM_MODULE_CODE_SUCCESS) {
        USER_LOG_ERROR("dji system init error\n");
    }

    /*----开始:在此处初始化需要实现的功能,下面以消息订阅初始化为例----*/

    /*-----------初始化消息订阅功能-------*/
    returnCode = DjiTest_FcSubscriptionStartService();
    if (returnCode != DJI_ERROR_SYSTEM_MODULE_CODE_SUCCESS) {
        USER_LOG_ERROR("data subscription init error\n");
    }

    /*----结束:在此处初始化需要实现的功能-------------------------*/

    /*-----------通知 PSDK 核心应用程序启动----------*/
    /*-----------需要在所有模块初始化和注册接口后完成调用----*/
    returnCode = DjiCore_ApplicationStart();
    if (returnCode != DJI_ERROR_SYSTEM_MODULE_CODE_SUCCESS) {
```
```

```
        USER_LOG_ERROR("start sdk application error");
    }
    /*-----------死循环，防止程序运行一次就退出---------------*/
    while (1) {
        sleep(1);
    }
    return 0;
}
```

1.4.7 构建项目

CMake 是一个跨平台、开源的构建系统，也是一个集软件构建、测试、打包于一体的软件。它使用独立于平台和编译器的配置文件对软件编译过程进行控制。CMake 通过 CMakeLists.txt 配置项目的构建系统，配合使用 cmake 命令行工具生成构建系统并执行编译、测试。

1. CMakeLists.txt 解析

```
# ---------CMakeLists.txt 在最开始需要声明 CMake 的最低版本要求--------
cmake_minimum_required(VERSION 3.5)

# ----------设置工程名-----------------------------------------
project(my_project_name)

# ----------设置编译选项---------------------------------------
set(CMAKE_C_FLAGS "-pthread -std=gnu99")
set(CMAKE_EXE_LINKER_FLAGS "-pthread")
set(CMAKE_C_COMPILER "gcc")
set(CMAKE_CXX_COMPILER "g++")
add_definitions(-D_GNU_SOURCE)
set(COMMON_CXX_FLAGS "-std=c++11 -pthread")
set(CMAKE_CXX_FLAGS "${CMAKE_CXX_FLAGS} -fprofile-arcs -ftest-coverage")
set(CMAKE_C_FLAGS "${CMAKE_C_FLAGS} -fprofile-arcs -ftest-coverage")
set(CMAKE_EXE_LINKER_FLAGS "${CMAKE_EXE_LINKER_FLAGS} -fprofile-arcs -ftest-coverage -lgcov")

#-------------声明值为 payloadsdk 的变量 PACKAGE_NAME--------------
set(PACKAGE_NAME payloadsdk)

# -----声明值为 arm-linux-gnueabihf-gcc 的交叉编译器变量 TOOLCHAIN_NAME--
set(TOOLCHAIN_NAME arm-linux-gnueabihf-gcc)

# ------设置树莓派开发板的硬件架构名称，连接树莓派，执行 uname -m 命令-------
add_definitions(-DPLATFORM_ARCH_armv7l=1)

# ----- file 命令用于指定需要访问文件系统中的文件和路径---------------------
```

```
file(GLOB_RECURSE MODULE_OSAL_SRC osal/*.c)
file(GLOB_RECURSE MODULE_HAL_SRC hal/*.c)
file(GLOB_RECURSE MODULE_3RDPART_SRC 3rdparty/*.c)
file(GLOB_RECURSE MODULE_MONITOR_SRC monitor/*.c)
file(GLOB_RECURSE MODULE_SAMPLE_SRC fc_subscription/*.c)

# ------ include_directories 命令用于在构建（Build）时添加包含目录。将指定的目录添加到
编译器（Compiler）中，用于搜索包含文件的目录------------------
include_directories(./)
include_directories(./osal)
include_directories(./hal)
include_directories(./3rdparty)
include_directories(./monitor)
include_directories(psdk_lib/include)
include_directories(./fc_subscription)

# ------ link_directories 命令用于指定第三方库所在的路径--------------------
link_directories(psdk_lib/lib/${TOOLCHAIN_NAME})
link_libraries(${CMAKE_CURRENT_LIST_DIR}/psdk_lib/lib/${TOOLCHAIN_NAME}/lib${PACKAGE_NAME}.a)

# --------设置 CMake 的二进制执行文件的路径-----------------------------
if (NOT EXECUTABLE_OUTPUT_PATH)
    set(EXECUTABLE_OUTPUT_PATH ${CMAKE_BINARY_DIR}/bin)
endif ()

# ------- 基于指定的源文件，创建可执行文件------------------------------
add_executable(${PROJECT_NAME}
        main.c
        dji_init.c
        ${MODULE_OSAL_SRC}
        ${MODULE_HAL_SRC}
        ${MODULE_3RDPART_SRC}
        ${MODULE_MONITOR_SRC}
        ${MODULE_SAMPLE_SRC})

# --------设置 3rdparty 的路径变量----------------------------------
set(CMAKE_MODULE_PATH ${CMAKE_CURRENT_SOURCE_DIR}/3rdparty)

# ----- find_package 命令用于查找并载入一个外部包----------------------------
find_package(LIBUSB REQUIRED)
if (LIBUSB_FOUND)
    message(STATUS "Found LIBUSB installed in the system")
    message(STATUS " - Includes: ${LIBUSB_INCLUDE_DIR}")
    message(STATUS " - Libraries: ${LIBUSB_LIBRARY}")
```

```
    # ------ add_definitions 命令用于添加编译选项--------------------
    add_definitions(-DLIBUSB_INSTALLED)

    # ------设置目标 PROJECT_NAME 需要的 usb-1.0 库--------------------
        target_link_libraries(${PROJECT_NAME} usb-1.0)
    else ()
        message(STATUS "Cannot Find LIBUSB")
    endif (LIBUSB_FOUND)

    target_link_libraries(${PROJECT_NAME} m)

    # -----为目标 PROJECT_NAME 添加规则，并通过执行命令生成输出结果。COMMAND 用于指定生成时要
    执行的命令行，WORKING_DIRECTORY 用于指定执行命令的路径-----------
    add_custom_command(TARGET ${PROJECT_NAME}
            PRE_LINK COMMAND cmake ..
            WORKING_DIRECTORY ${CMAKE_BINARY_DIR})
```

2 编译项目

进入代码的目录，使用如下命令将代码编译为程序。

```
mkdir build
cd build
cmake ..
make
```

执行 C 语言程序，使用 cd build/bin/命令进入程序的目录，使用 sudo ./my_project_name 命令运行示例程序。

● 任务设计

在了解 DJI 无人机的相关概念，熟悉 PSDK 基础知识、工作原理和开发基础后，我们将搭建 PSDK 的开发环境。

任务的具体设计要求如下。

（1）准备无人机、PSDK 拓展接口板、树莓派开发板。

（2）准备串口线、网线。

（3）通过串口线连接树莓派开发板。

（4）配置树莓派无线网络。

（5）开启树莓派 SSH。

（6）按照 1.4.1 节中的通用结构构建项目的基础代码。

● 任务实施

任务的实施步骤和结果如下。

（1）在教师的指导下连接设备。

（2）在教师的指导下使用串口线连接树莓派开发板。

（3）配置树莓派无线网络。

执行 ls /sys/class/net/ 命令查看树莓派的 Wi-Fi，通过确认设备节点判断是否为硬件故障或没有加载驱动。

```
pi@raspberrypi:~ $ ls /sys/class/net/
eth0 lo wlan0 ztrtavngka
pi@raspberryp1:~
pi@raspberrypi:~$ls /sys/class/net/
ethe0 lo wlan0 ztrtavngka
pi@raspberrypi:~ $
```

说明：结果中显示有 wlan0，表示树莓派 Wi-Fi 设备已经被识别且无线网卡驱动已加载。

在 raspi-config 的首页中，执行 sudo iw dev wlan0 scan | grep SSID 命令确认搜索到 Wi-Fi，如图 1-4 所示。

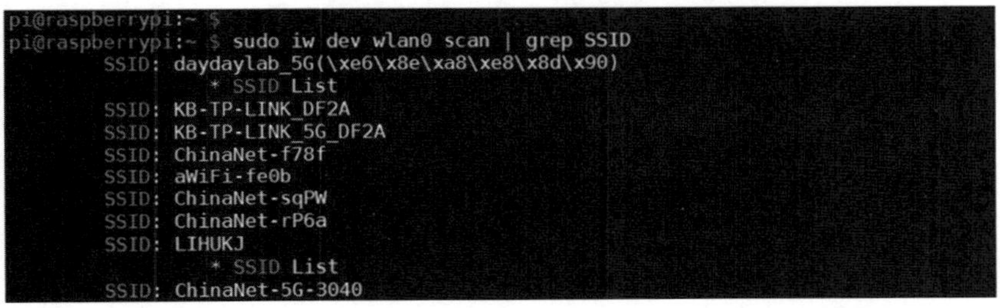

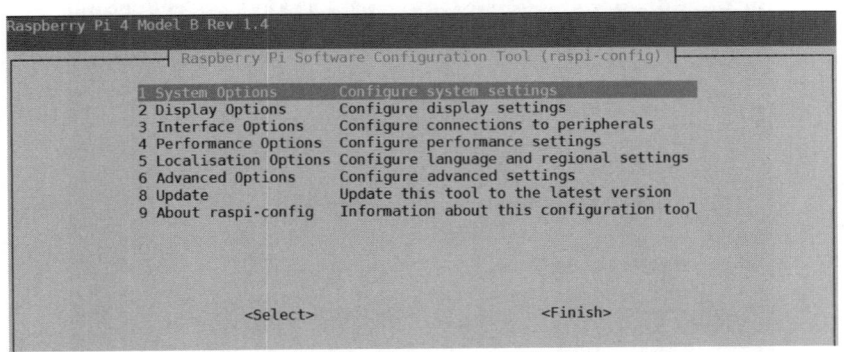

图 1-4　确认搜索到 Wi-Fi

通过命令行的方式配置无线网络，执行 sudo raspi-config 命令打开树莓派配置界面，如图 1-5 所示。

图 1-5　树莓派配置界面

选择"1 System Options　Configure system settings"选项，进入"Raspberry PI Software Configuration Tool（raspi-config）"界面，选择"S1 Wireless LAN　Enter SSID and passphrase"选项。选择 WLAN，如图 1-6 所示。在跳转的界面中输入 WLAN 的 SSID，如图 1-7 所示。

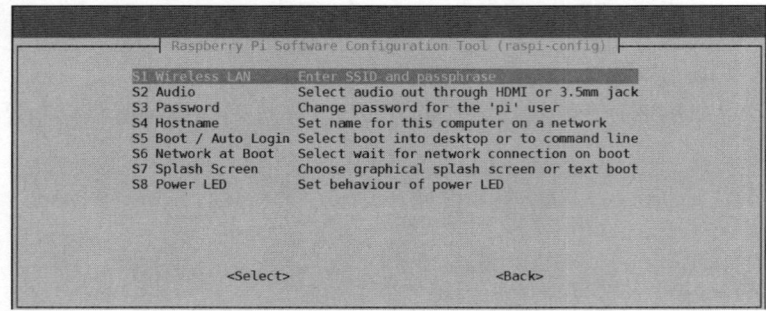

图 1-6　选择 WLAN

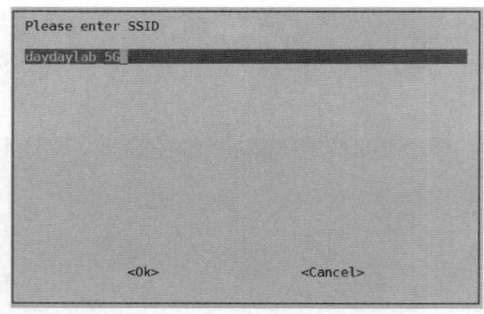

图 1-7　输入 WALN 的 SSID

输入密码并按"Enter"键，回到 raspi-config 的首页。按"Tab"键选择"Finish"选项并重启，此时在终端中输入"ifconfig"，正常情况下能看到 wlan0 动态获取的 IP 地址，如图 1-8 所示。

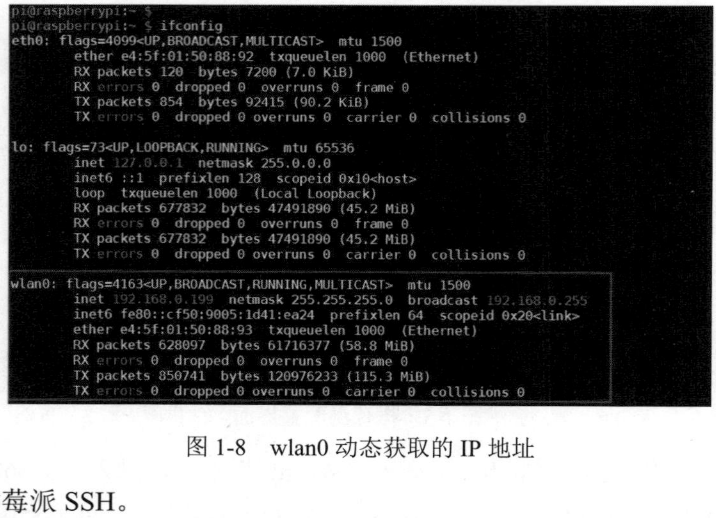

图 1-8　wlan0 动态获取的 IP 地址

（4）开启树莓派 SSH。

在终端中输入"sudo raspi-config"打开树莓派配置界面，选择"3 Interface Options Configure connections to peripherals"选项，如图 1-9 所示。在跳转的界面中选择"I2 SSH Enable/disable remote command line access using SSH"选项，如图 1-10 所示，配置 SSH。

项目 1　无人机与 PSDK 开发概述

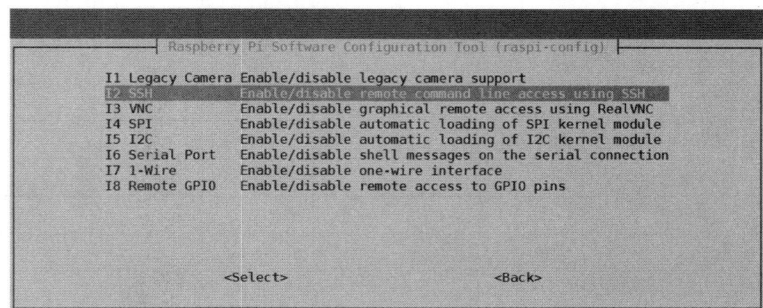

图 1-9　选择"3 Interface Options　Configure connections to peripherals"选项

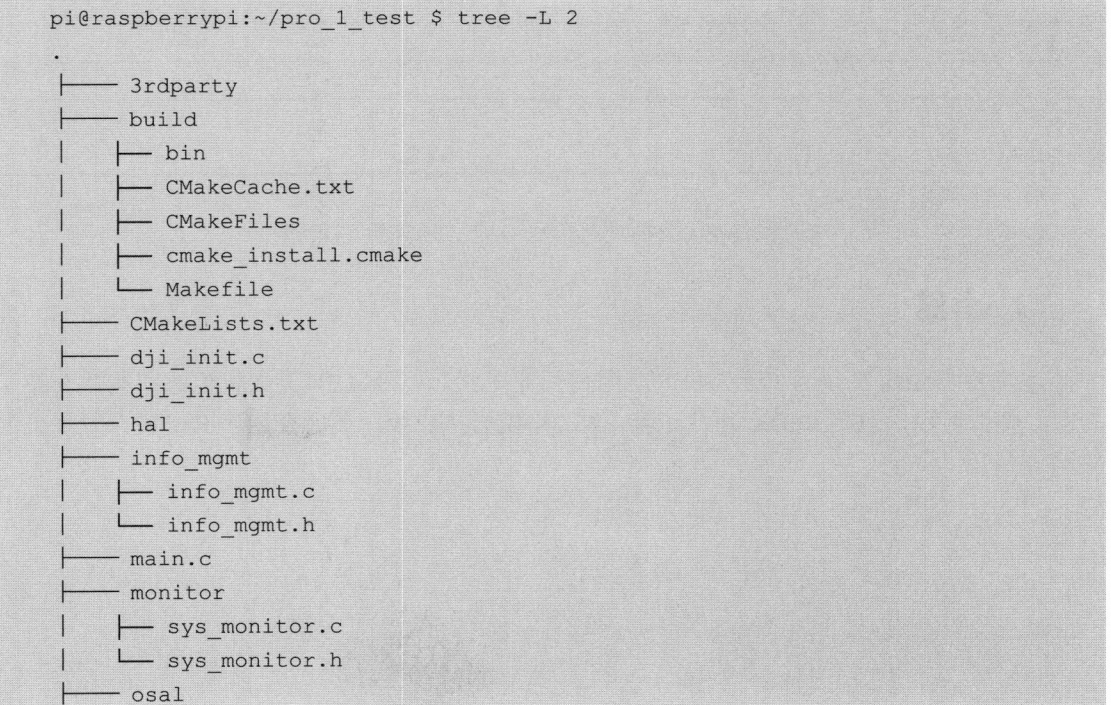

图 1-10　选择"I2 SSH　Enable/disable remote command line access using SSH"选项

(5) 按照 1.4.1 节中的通用结构构建项目的基础代码。

在教师的指导下完成第一个项目的基础代码，项目名称为 **pro_1_test**，目录结构如下。

```
pi@raspberrypi:~/pro_1_test $ tree -L 2
.
├── 3rdparty
├── build
│   ├── bin
│   ├── CMakeCache.txt
│   ├── CMakeFiles
│   ├── cmake_install.cmake
│   └── Makefile
├── CMakeLists.txt
├── dji_init.c
├── dji_init.h
├── hal
├── info_mgmt
│   ├── info_mgmt.c
│   └── info_mgmt.h
├── main.c
├── monitor
│   ├── sys_monitor.c
│   └── sys_monitor.h
├── osal
```

25

```
└── psdk_lib
    ├── include
    └── lib
```

任务评价

任务过程评价表

任务实施人姓名_____ 学号_____ 时间_____

评价项目及标准		分值	小组评议	教师评议
技术能力	1. 熟悉基本概念	10		
	2. 了解无人机的应用与行业现状	10		
	3. 熟悉 PSDK 拓展接口板	10		
	4. 掌握 PSDK 开发基础	10		
	5. 熟悉 PSDK 开发环境的搭建方法	10		
	6. 掌握项目开发的通用结构	10		
执行能力	1. 出勤情况	5		
	2. 遵守纪律情况	5		
	3. 是否主动参与，有无提问记录	5		
	4. 有无职业意识	5		
社会能力	1. 能否有效沟通	5		
	2. 是否使用基本的文明礼貌用语	5		
	3. 能否与组员主动交流、积极合作	5		
	4. 能否自我学习、自我管理	5		
总分		100		
评定等级：				
评价意见		学习意见		

评定等级具体如下。

A：优，得分>90。B：好，90≥得分>80。C：一般，80≥得分>60。D：有待提高，得分≤60。

小结

本项目首先简要介绍了无人机的应用与行业现状，以及 DJI 无人机生态；其次介绍了 PSDK 硬件环境，包括 PSDK 开发环境、PSDK 拓展接口板；最后介绍了 PSDK 开发基础的内容。

第 2 篇
PSDK 功能模块

- 项目 2　日志管理与电源管理
- 项目 3　无人机信息管理
- 项目 4　相机管理与云台管理
- 项目 5　飞行控制与运动规划
- 项目 6　自定义控件
- 项目 7　时间同步与精准定位
- 项目 8　SDK 互联互通
- 项目 9　视频流传输与回放下载
- 项目 10　健康管理系统（HMS）
- 项目 11　获取相机码流（liveview）
- 项目 12　本地升级
- 项目 13　相机实现与云台实现

项目 2 日志管理与电源管理

🔴 任务要求

本项目将使用 DJI PSDK API 进行电源管理，同时使用日志管理功能将电源管理的日志输出到控制台中。

🔴 知识导入

2.1 日志管理

PSDK 的日志管理功能支持通过串口、终端或 USB 等日志输出方法，输出 Debug、Info、Warn 和 Error 四种等级的日志；使用能够显示日志颜色的终端工具，如 XShell、SecureCRT、Putty 等，可以不同的颜色显示不同等级的日志。

2.1.1 日志管理功能

基于 PSDK 开发的负载设备输出的日志结构：日志颜色起始标识符+系统时间+模块名称+日志等级标识+日志内容+日志颜色结束标识符。

1. 日志的颜色标识符

当使用不支持显示日志颜色的调试工具查看日志时，会显示日志的颜色标识符。

（1）日志颜色：不同等级的日志，其标识颜色也不同。XShell、SecureCRT、Putty 等工具能够根据日志等级，以不同的颜色显示日志。

（2）日志颜色起始标识符。

- \033[30m（黑色）、\033[31m（红色）、\033[32m（绿色）。

- \033[33m（黄色）、\033[34m（蓝色）、\033[35m（紫色）。
- \033[36m（青色）、\033[37m（白色）。

（3）日志颜色结束标识符：\033[0m。

2. 日志信息

（1）系统时间：在负载设备上电时，负载设备的时间为负载设备系统的时间；当负载设备与无人机完成时间同步后，负载设备的时间将与无人机的时间同步。

（2）模块名称：PSDK 模块的名称。用户打印接口的模块名称为 user。

（3）日志等级：日志等级从高到低为 Debug、Info、Warn 和 Error，日志管理功能模块可打印不高于指定等级的所有日志。日志等级如表 2-1 所示。

表 2-1　日志等级

日志等级	日志内容	输出接口	日志颜色
Debug - 4	调试信息	USER_LOG_DEBUG	White
Info - 3	关键信息	USER_LOG_INFO	Green
Warn - 2	警告信息	USER_LOG_WARN	Yellow
Error - 1	系统错误	USER_LOG_ERROR	Red

（4）日志内容：单条日志最多不超过 500 字节（Bytes）。

2.1.2　接口说明

日志管理相关功能的头文件为 dji_logger.h，该文件描述了日志管理功能中结构体和函数原型的关键信息与使用方法。

1. 头文件中的宏定义、枚举与结构体

- ConsoleFunc：需要被注册的控制台方法。
- E_DjiLoggerConsoleLogLevel：日志级别。
- T_DjiLoggerConsole：日志等级。

具体说明如下。

1）ConsoleFunc 新类型

日志管理功能模块通过 typedef 定义了 ConsoleFunc 指针函数，该函数用于表示需要被注册的控制台方法。ConsoleFunc 指针函数的定义如下。

```
typedef T_DjiReturnCode (*ConsoleFunc)(const uint8_t *data, uint16_t dataLen);
```

说明：typedef 的定义形式为"typedef 返回类型 (*新类型)(参数表)"。typedef 的功能是定义新的类型。

2）日志级别

日志管理功能模块通过枚举类型 E_DjiLoggerConsoleLogLevel 定义了 4 个日志等级。

```
typedef enum {
    DJI_LOGGER_CONSOLE_LOG_LEVEL_ERROR = 0,
```

```
    DJI_LOGGER_CONSOLE_LOG_LEVEL_WARN = 1,
    DJI_LOGGER_CONSOLE_LOG_LEVEL_INFO = 2,
    DJI_LOGGER_CONSOLE_LOG_LEVEL_DEBUG = 3,
} E_DjiLoggerConsoleLogLevel;
```

各日志等级的功能如下。

（1）DJI_LOGGER_CONSOLE_LOG_LEVEL_ERROR = 0。

打印系统错误类型（Error）的日志。控制台的方法与等级是彼此关联的。如果注册的控制台方法的级别低于当前等级，则等级接口不会被成功打印。

（2）DJI_LOGGER_CONSOLE_LOG_LEVEL_WARN = 1。

打印警告信息类型（Warn）的日志。控制台的方法与等级是彼此关联的。如果注册的控制台方法的级别低于当前等级，则等级接口不会被成功打印。

（3）DJI_LOGGER_CONSOLE_LOG_LEVEL_INFO = 2。

打印关键信息类型（Info）的日志。控制台的方法与等级是彼此关联的。如果注册的控制台方法的级别低于当前等级，则等级接口不会被成功打印。

（4）DJI_LOGGER_CONSOLE_LOG_LEVEL_DEBUG = 3。

打印调试信息类型（Debug）的日志。控制台的方法与等级是彼此关联的。如果注册的控制台方法的级别低于当前等级，则等级接口不会被成功打印。

3）日志等级

日志管理功能模块通过结构体 T_DjiLoggerConsole 定义日志等级。

```
typedef struct {
    uint8_t consoleLevel;
    ConsoleFunc func;
    bool isSupportColor;
} T_DjiLoggerConsole;
```

说明：参数 consoleLevel 指定所需要打印的日志等级；参数 func 指定日志输出方式，在注册日志输出方式前，要先测试该日志输出方式是否能够正常打印用户所需要的日志。

2. 函数原型

- T_DjiReturnCode DjiLogger_AddConsole：添加控制台功能与日志等级。
- T_DjiReturnCode DjiLogger_RemoveConsole：去除控制台功能与日志等级。
- void DjiLogger_UserLogOutput：打印特定格式的选定等级的日志。

具体说明如下。

1）添加控制台功能与日志等级

PSDK 使用 DjiLogger_AddConsole 函数添加控制台功能与日志等级。在注册控制台前，用户需要提供控制台方法及与该方法相关的日志等级。日志管理功能模块可以打印所有不高于特定级别的日志。最多支持同时注册 8 种不同的控制台方法。在注册控制台方法前，需要测试注册模块，以确保所有的方法都是正常的。如果用户同时注册了多个模块，则所有的方法都会被打印。

DjiLogger_AddConsole 函数定义如下。

```
T_DjiReturnCode DjiLogger_AddConsole(T_DjiLoggerConsole *console);
```
说明：参数 console 指向控制台方法。该函数根据程序的执行情况输出对应的返回值。

2）去除控制台功能与日志等级

PSDK 使用 DjiLogger_RemoveConsole 函数去除控制台功能与日志等级，该函数定义如下。

```
T_DjiReturnCode DjiLogger_RemoveConsole(T_DjiLoggerConsole *console);
```
说明：参数 console 指向控制台方法。该函数根据程序的执行情况输出对应的返回值。

3）打印特定格式的选定等级的日志

PSDK 使用 DjiLogger_UserLogOutput 函数通过注册方法打印特定格式的选定等级的日志。注册的方法是按照对应级别打印的。如果控制台的级别比需要打印的日志的等级低，则不会打印成功。

DjiLogger_UserLogOutput 函数定义如下。

```
void DjiLogger_UserLogOutput(E_DjiLoggerConsoleLogLevel level, const char *fmt, ...);
```
说明：参数 level 指定所需要打印的日志等级；参数 fmt 指向用户需要打印的日志内容。

3. 日志打印预处理

dji_logger.h 头文件通过#define 定义了 4 条预处理命令：USER_LOG_DEBUG、USER_LOG_INFO、USER_LOG_WARN 和 USER_LOG_ERROR。

通过封装 DjiLogger_UserLogOutput 函数，使日志打印的调用更加方便。预处理命令如下。

```
#define USER_LOG_DEBUG(fmt, ...) \
        DjiLogger_UserLogOutput(DJI_LOGGER_CONSOLE_LOG_LEVEL_DEBUG, "[%s:%d) " fmt, __FUNCTION__, __LINE__ , ##__VA_ARGS__)
#define USER_LOG_INFO(fmt, ...) \
        DjiLogger_UserLogOutput(DJI_LOGGER_CONSOLE_LOG_LEVEL_INFO, "[%s:%d) " fmt, __FUNCTION__, __LINE__ , ##__VA_ARGS__)
#define USER_LOG_WARN(fmt, ...) \
        DjiLogger_UserLogOutput(DJI_LOGGER_CONSOLE_LOG_LEVEL_WARN, "[%s:%d) " fmt, __FUNCTION__, __LINE__ , ##__VA_ARGS__)
#define USER_LOG_ERROR(fmt, ...) \
        DjiLogger_UserLogOutput(DJI_LOGGER_CONSOLE_LOG_LEVEL_ERROR, "[%s:%d) " fmt, __FUNCTION__, __LINE__ , ##__VA_ARGS__)
```

说明：凡是以"#"开头的语句均为预处理命令，"define"为宏定义命令，"标识符"为所定义的宏名。

2.2 电源管理

将基于 PSDK 开发的负载设备安装在无人机的云台上，负载设备通过功率申请能获得较高的功率，因此，使用高功率的负载设备必须同时支持使用低功率；为了防止负载设备因未执行关机准备而丢失数据或意外损毁，PSDK 还提供了下发关机通知的功能。

2.2.1 电源管理功能

基于 PSDK 开发的负载设备使用电源管理功能可以实现高功率申请和关机通知。

1. 高功率申请

负载设备能够向无人机申请高功率。高功率申请完成后，可使用电压表等仪器测量 SkyPort v2 和 X-PORT 指定接口的电压。在不使用高功率申请时，PSDK 默认功率输出电压为 13.6V，申请高功率之后的电压为 17V。

2. 关机通知

当手动点按-长按无人机的电源开关后，无人机接收关机通知准备关机，同时向基于 PSDK 开发的负载设备发送关机通知。在电源管理中注册了该通知的回调函数后，负载设备可以完成关机前需要执行的操作，如保存数据等。

无人机下发关机通知的流程如下。
- 当无人机获取用户发送的关机通知后，将向基于 PSDK 开发的负载设备发送关机通知。
- 当负载设备接收到无人机发送的关机通知后，将完成关机前需要执行的操作，并更改关机准备状态。
- 当无人机获取所有负载设备关机准备的状态后，将会立即关机。

需要注意的是，无人机在接收到关机通知的一段时间后，将会强制关机，具体强制关机的时间参见对应产品型号的说明书。

2.2.2 接口说明

PSDK 电源管理相关功能的头文件为 psdk_lib/include/dji_power_management.h，该文件描述了电源管理功能中结构体和函数原型的关键信息与使用方法。

1. 头文件中的宏定义、枚举与结构体

- E_DjiPowerManagementPinState：高功率引脚的状态。
- DjiWriteHighPowerApplyPinCallback：设置高功率引脚的状态。
- DjiPowerOffNotificationCallback：设置关机通知。

具体说明如下。

1）高功率引脚的状态

通过枚举类型 E_DjiPowerManagementPinState 设置高功率引脚的状态。

```
typedef enum {
    DJI_POWER_MANAGEMENT_PIN_STATE_RESET = 0,    # 引脚为低电压
    DJI_POWER_MANAGEMENT_PIN_STATE_SET = 1,      # 引脚为高电压
} E_DjiPowerManagementPinState;
```

说明：需要在树莓派开发板和无人机扩展板之间增加一条功率控制引脚连线。

2）设置高功率引脚状态的回调函数

通过 typedef 定义了 DjiWriteHighPowerApplyPinCallback 回调函数，该函数用于设置高功率引脚的状态。

```
typedef T_DjiReturnCode (*DjiWriteHighPowerApplyPinCallback)
(E_DjiPowerManagementPinState pinState);
```

说明：通过 typedef 定义了 DjiWriteHighPowerApplyPinCallback 回调函数，该函数将 pinState 作为参数并返回 T_DjiReturnCode 类型。pinState 为 E_DjiPowerManagementPinState 中定义的高功率引脚的状态。

3）设置关机通知的回调函数

通过 typedef 定义了 DjiPowerOffNotificationCallback 回调函数，该函数用于通知负载设备无人机将要关机并获取负载设备关机准备状态。

```
typedef T_DjiReturnCode (*DjiPowerOffNotificationCallback)(bool
*powerOffPreparationFlag);
```

说明：通过 typedef 定义了 DjiPowerOffNotificationCallback 回调函数，该函数将布尔值 powerOffPreparationFlag 作为参数并返回 T_DjiReturnCode 类型。

2. 头文件函数原型

- DjiPowerManagement_Init：初始化电源管理模块。
- DjiPowerManagement_ApplyHighPowerSync：从无人机向负载设备施加高功率。
- DjiPowerManagement_RegWriteHighPowerApplyPinCallback：设置高功率应用引脚的电平状态的注册回调函数。
- DjiPowerManagement_RegPowerOffNotificationCallback：通知 Payload 无人机即将关机，同时获取 Payload 是否准备好关机状态的注册回调函数。

具体说明如下。

1）模块初始化

PSDK 使用 DjiPowerManagement_Init 函数初始化电源管理模块，用户在使用电源管理功能前应调用该函数。

DjiPowerManagement_Init 函数的定义如下。

```
T_DjiReturnCode DjiPowerManagement_Init(void);
```

2）施加高功率

PSDK 使用 DjiPowerManagement_ApplyHighPowerSync 函数在阻塞模式下从无人机向负载设备施加高功率。

在调用该函数前，需先使用 DjiPowerManagement_RegWriteHighPowerApplyPinCallback 函数注册回调函数，用于设置高功率应用引脚的电平状态。施加高功率后，DJI 适配器的电源引脚将根据预定规格输出高功率。该函数的最大执行时间略大于 600ms。

DjiPowerManagement_ApplyHighPowerSync 函数定义如下。

```
T_DjiReturnCode DjiPowerManagement_ApplyHighPowerSync(void);
```

3）设置高功率回调函数

PSDK 使用 DjiPowerManagement_RegWriteHighPowerApplyPinCallback 函数注册回调函数，用于设置高功率应用引脚的电平状态。

DjiPowerManagement_RegWriteHighPowerApplyPinCallback 函数定义如下。

```
T_DjiReturnCode DjiPowerManagement_RegWriteHighPowerApplyPinCallback(
DjiWriteHighPowerApplyPinCallback callback);
```

说明：参数 callback 是一个指向回调函数的指针。

4）设置关机回调函数

PSDK 使用 DjiPowerManagement_RegPowerOffNotificationCallback 函数注册回调函数，让无人机通知负载设备无人机即将关机并确认负载设备是否准备好关机状态。

注册完成后，负载设备会在无人机关机时以固定频率调用回调函数。用户需填写关机准备标志，一旦负载设备准备关机，回调函数将不会被调用。收到无人机端的下电通知后，设备端需要在 3s～5s 内完成关机准备。

DjiPowerManagement_RegPowerOffNotificationCallback 函数定义如下。

```
T_DjiReturnCode DjiPowerManagement_RegPowerOffNotificationCallback(
DjiPowerOffNotificationCallback callback);
```

说明：参数 callback 是一个指向回调函数的指针，用于通知无人机的关机消息并获取有效载荷关机准备标志。

2.2.3　树莓派 GPIO

树莓派的开发功能十分强大，它提供了一组用于输入与输出的脚针，称为 GPIO（Gereral-Purpose Input/Output）。

GPIO 引脚既可以用于输出高/低电平，又可以用于读入引脚的状态。GPIO 是一个比较重要的概念，用户可以通过 GPIO 引脚和硬件进行数据交互（如 UART）、控制硬件工作（如 LED、蜂鸣器等）、读取硬件的工作状态信号（如中断信号）等。

1. 查看 GPIO 版本

输入"gpio -v"查看树莓派内置的版本。

```
pi@raspberrypi:~ $ gpio -v
gpio version: 2.52
Copyright (c) 2012-2018 Gordon Henderson
This is free software with ABSOLUTELY NO WARRANTY.
For details type: gpio -warranty

Raspberry Pi Details:
  Type: Pi 4B, Revision: 04, Memory: 8192MB, Maker: Sony
  * Device tree is enabled.
  *--> Raspberry Pi 4 Model B Rev 1.4
  * This Raspberry Pi supports user-level GPIO access.
pi@raspberrypi:~ $
```

2. GPIO 引脚

目前主流的树莓派型号通常都有 40 个 GPIO 引脚可供使用。GPIO 引脚如图 2-1 所示。

wiringPi 编码	BCM 编码	功能名	物理引脚 BOARD 编码		功能名	BCM 编码	wiringPi 编码
		3.3V	1	2	5V		
8	2	SDA.1	3	4	5V		
9	3	SCL.1	5	6	GND		
7	4	GPIO.7	7	8	TXD	14	15
		GND	9	10	RXD	15	16
0	17	GPIO.0	11	12	GPIO.1	18	1
2	27	GPIO.2	13	14	GND		
3	22	GPIO.3	15	16	GPIO.4	23	4
		3.3V	17	18	GPIO.5	24	5
12	10	MOSI	19	20	GND		
13	9	MISO	21	22	GPIO.6	25	6
14	11	SCLK	23	24	CE0	8	10
		GND	25	26	CE1	7	11
30	0	SDA.0	27	28	SCL.0	1	31
21	5	GPIO.21	29	30	GND		
22	6	GPIO.22	31	32	GPIO.26	12	26
23	13	GPIO.23	33	34	GND		
24	19	GPIO.24	35	36	GPIO.27	16	27
25	26	GPIO.25	37	38	GPIO.28	20	28
		GND	39	40	GPIO.29	21	29

图 2-1　GPIO 引脚

在这 40 个引脚中，除了 12 个电源类的 GPIO 引脚（5V、3.3V 和 GND 引脚），其余 28 个都是可编程的 GPIO 引脚，其中，部分 GPIO 引脚可以复用为 I2C（集成电路总线）、SPI（串行外设接口）、UART（通用异步收发传输器）、PWM 等，以用来驱动各种外设。

3. wiringPi 库

wiringPi 库是应用于树莓派平台的 GPIO 控制库函数，使用 C 语言开发，提供了丰富的接口，如 GPIO 控制、中断、多线程等。

在使用 wiringPi 库时，需要进行如下操作。

1）包含头文件

```
#include<wiringPi.h>
```

2）初始化树莓派

在执行任何操作前都必须先初始化树莓派，否则程序不能正常工作。硬件初始化的函数原型如下。

```
int wiringPiSetup (void)
```

说明：当使用这个函数初始化树莓派时，程序使用的是 wiringPi 引脚编号表。返回执行状态，-1 表示初始化失败。

3）配置引脚 I/O 模式

GPIO 设置引脚 I/O 模式的函数原型如下。

```
void pinMode (int pin, int mode)
```

说明：参数 pin 表示配置的引脚；参数 mode 指定引脚 I/O 模式，可取的值有 INPUT、OUTPUT、PWM_OUTPUT、GPIO_CLOCK。

4）输出指定的电平信号

让一个已经被配置为输出模式的引脚输出指定的电平信号，函数原型如下。

```
void digitalWrite (int pin, int value)
```

说明：参数 pin 表示控制的引脚；参数 value 表示引脚输出的电平值，可取的值有 HIGH 和 LOW，分别代表高电平和低电平。

5）读取一个引脚的电平值

读取一个引脚的电平值的函数原型如下。

```
int digitalRead (int pin)
```

说明：参数 pin 表示读取的引脚；返回值是引脚上的电平，可以是 LOW 或 HIGH。

更多关于 wiringPi 库的函数和方法，请参阅树莓派官网。

2.3 代码模板

当开发者实现电源管理功能时，不要在 main 函数中调用电源管理接口，而需要创建用户线程来实现电源管理功能。在用户线程中调用电源管理接口，启动调度器后，电源管理接口将正常运行。本项目的模板文件位于 pro_2_source_template 目录中。

1. 代码结构

下面是实现电源管理功能的代码框架，用户只需要编写用于实现电源管理功能的 power_mgmt.c 和 power_mgmt.h 即可。

```
├── 3rdparty              // 第三方库文件目录，省略了该目录下的内容
├── CMakeLists.txt        // cmake 构建项目的文件
├── dji_init.c            // 系统初始化源文件
├── dji_init.h            // 系统初始化头文件
├── power_mgmt            // 用户实现电源管理功能的目录
│   ├── power_mgmt.c      // 实现电源管理功能的源文件
│   └── power_mgmt.h      // 头文件
├── hal                   // hal 文件目录，省略了该目录下的内容
├── main.c                // 主函数文件
├── monitor               // 监控文件目录，省略了该目录下的内容
├── osal                  // osal 文件目录，省略了该目录下的内容
└── psdk_lib              // psdk 库文件目录，省略了该目录下的内容
```

2. 头文件模板

本项目的头文件名为 power_mgmt.h。该头文件定义的内容可满足正常的应用开发需求，如果不需要实现特殊的功能，则无须修改该文件。power_mgmt.h 中定义了两个函数：DjiTest_PowerManagementStartService 和 DjiTest_RegApplyHighPowerHandler。

3. 源文件模板

本项目的源文件名为 power_mgmt.c，以下是 power_mgmt.c 的模板文件，按照提示即可完成代码的编写。

```c
#include "power_mgmt.h"
#include "dji_logger.h"
#include "dji_aircraft_info.h"

static T_DjiReturnCode DjiTest_PowerOffNotificationCallback(bool *powerOffPreparationFlag);

static T_DjiTestApplyHighPowerHandler s_applyHighPowerHandler;

T_DjiReturnCode DjiTest_RegApplyHighPowerHandler(T_DjiTestApplyHighPowerHandler *applyHighPowerHandler)
{
    if (applyHighPowerHandler->pinInit == NULL) {
        USER_LOG_ERROR("reg apply high power handler pinInit error");
        return DJI_ERROR_SYSTEM_MODULE_CODE_INVALID_PARAMETER;
    }

    if (applyHighPowerHandler->pinWrite == NULL) {
        USER_LOG_ERROR("reg apply high power handler pinWrite error");
        return DJI_ERROR_SYSTEM_MODULE_CODE_INVALID_PARAMETER;
    }

    memcpy(&s_applyHighPowerHandler, applyHighPowerHandler, sizeof(T_DjiTestApplyHighPowerHandler));

    return DJI_ERROR_SYSTEM_MODULE_CODE_SUCCESS;
}

T_DjiReturnCode DjiTest_PowerManagementStartService(void)
{
    T_DjiReturnCode returnCode;
    T_DjiAircraftInfoBaseInfo baseInfo = {0};

    /* ---------------在此处完成以下功能-------------------------- */
    /* step 1 调用 DjiPowerManagement_Init 函数完成电源管理模块的初始化*/
    /* step 2 调用 DjiAircraftInfo_GetBaseInfo 函数获取无人机的参数*/
    /* step 3 负载设备申请高功率*/
    /* step 3.1 获取树莓派初始化状态*/
    /* step 3.2 获取当前高电平写入状态*/
    /* step 3.3 初始化树莓派*/
    /* step 3.4 通过回调函数设置高功率申请*/
    /* step 3.5 负载设备能够向无人机申请高功率*/
```

```
    /* step 4 注册关机通知的回调函数*/
    /* step 5 实现关机通知的回调函数的关机功能*/

    /* ----------------------------------------------------------- */

    /* step 1 调用 DjiPowerManagement_Init 函数完成电源管理模块的初始化---- */

    /* step 2 调用 DjiAircraftInfo_GetBaseInfo 函数获取无人机的参数----*/

    /* step 3 完成负载设备申请高功率 ---------------------------------*/
    if (baseInfo.aircraftType == DJI_AIRCRAFT_TYPE_M300_RTK &&
        (baseInfo.djiAdapterType == DJI_SDK_ADAPTER_TYPE_SKYPORT_V2 ||
         baseInfo.djiAdapterType == DJI_SDK_ADAPTER_TYPE_XPORT)) {

        // step 3.1 获取当前树莓派初始化状态 s_applyHighPowerHandler.pinInit

        // step 3.2 获取当前高电平写入状态 s_applyHighPowerHandler.pinWrite

        // step 3.3 在此处初始化树莓派

        // step 3.4  在此处通过DjiPowerManagement_RegWriteHighPowerApplyPinCallback
回调函数设置高功率申请

        // step 3.5 在此处通过调用 DjiPowerManagement_ApplyHighPowerSync 函数实现负
载设备能够向无人机申请高功率的功能

    }

    /* step 4 注册关机通知的 DjiPowerManagement_RegPowerOffNotificationCallback
回调函数---------------*/

    return DJI_ERROR_SYSTEM_MODULE_CODE_SUCCESS;
}

/* step 5 在此处实现关机通知的回调函数的关机功能--------------------*/
static T_DjiReturnCode DjiTest_PowerOffNotificationCallback(bool
*powerOffPreparationFlag)
{
    USER_LOG_INFO("aircraft will power off soon.");
    // 添加关机功能

    *powerOffPreparationFlag = true;

    return DJI_ERROR_SYSTEM_MODULE_CODE_SUCCESS;
}
```

任务设计

电源管理将实现负载设备的高功率申请和负载设备的关机通知功能，同时通过日志管理功能将信息输出到控制台中。

任务的具体设计要求如下。

（1）创建项目目录。

（2）复制 psdk_lib、hal、osal、monitor、3rdparty、main.c 和 CMakeLists.txt 到该项目目录下。

（3）本项目将实现电源管理功能，项目目录下的 power_mgmt 目录用于编写代码，以实现用户线程调用电源管理 API 接口。

（4）在 power_mgmt 目录下编写代码，实现电源管理。

① 初始化电源管理功能。

② 获取无人机的参数。

③ 初始化树莓派。

④ 通过回调函数设置高功率申请。

⑤ 实现负载设备能够向无人机申请高功率的功能。

⑥ 注册关机通知的回调函数。

⑦ 实现关机通知的回调函数的关机功能。

（5）使用 cmake 编译项目。

（6）执行项目。

① 查看高功率。

② 查看负载设备关机。

任务实施

基于 PSDK 实现电源管理的负载设备高功率申请和关机通知，同时使用日志管理功能将日志输出到控制台中。限于篇幅，这里只给出任务实施步骤和关键代码。

1. 模板文件

连接树莓派开发板，默认路径为当前登录用户的家目录。查看当前的家目录，该目录下有一个名为 pro_2_source_template 的目录，此目录下保存了本项目的所有模板文件。

目录结构如下。

```
pi@raspberrypi:~ $ cd pro_2_source_template/
pi@raspberrypi:~/pro_2_source_template $ tree
.
├── 3rdparty              # 为目录，省略了目录下的文件
├── CMakeLists.txt        # cmake 构建项目文件
├── dji_init.c            # 系统初始化源文件
├── dji_init.h            # 系统初始化头文件
├── hal                   # 为目录，省略了目录下的文件
├── power_mgmt            # 电源管理功能的目录
```

```
        |      ├── power_mgmt.c          # 电源管理功能的源代码
        |      └── power_mgmt.h          # 头文件
        ├── main.c                       # 主函数文件
        ├── monitor                      # 为目录，省略了目录下的文件
        ├── osal                         # 为目录，省略了目录下的文件
        └── psdk_lib                     # 为目录，省略了目录下的文件
```

2. 复制模板目录

将 pro_2_source_template 目录下的所有文件和目录复制到 pro_2_power_mgmt 目录中。

```
pi@raspberrypi:~ $ cp -r pro_2_source_template pro_2_power_mgmt
```

3. 修改 dji_init.c

1）添加日志管理头文件

本项目涉及日志管理功能，需要在 dji_init.c 中添加日志管理头文件。

```
/* ---------在此处添加日志管理头文件 dji_logger.h ---------------------*/
#include <dji_logger.h>
/* ------------------------------------------------------------------*/
```

2）设置日志存放路径

```
/*---------在此处设置存放日志的目录、路径及日志文件名索引----------------*/
#define DJI_LOG_PATH                    "Logs/DJI"
#define DJI_LOG_INDEX_FILE_NAME         "Logs/latest"
#define DJI_LOG_FOLDER_NAME             "Logs"
/* ------------------------------------------------------------------*/
```

3）填写用户账号信息

在 dji_init.c 中填写用户账号信息，开发者需根据实际的 PSDK 用户参数填写。

```
#define USER_App_NAME                   "RaspberryPi"
#define USER_App_ID                     "126413"
#define USER_App_KEY                    "0ea2eadb314d1a6e870975364e2bf4a"
#define USER_App_LICENSE                "HZ/GptycWQ........................"
#define USER_DEVELOPER_ACCOUNT          "abc@outlo**.com"
#define USER_BAUD_RATE                  "460800"
```

4）添加日志管理功能

在 static T_DjiReturnCode DjiUser_PrepareSystemEnvironment 函数中添加以下日志管理功能。

```
static T_DjiReturnCode DjiUser_PrepareSystemEnvironment(void)
{
    /* step 1 设置控制台输出参数-----------*/
    T_DjiLoggerConsole printConsole = {
        .func = DjiUser_PrintConsole,
        .consoleLevel = DJI_LOGGER_CONSOLE_LOG_LEVEL_INFO,
        .isSupportColor = true,
    };
```

```
    /* --------------------------------*/

    /* step 2 设置日志文件输出参数---------*/
    T_DjiLoggerConsole localRecordConsole = {
        .consoleLevel = DJI_LOGGER_CONSOLE_LOG_LEVEL_DEBUG,
        .func = DjiUser_LocalWrite,
        .isSupportColor = true,
    };
    /* --------------------------------*/

    /* step 3 设置日志系统文件位置-----------*/
    if (DjiUser_LocalWriteFsInit(DJI_LOG_PATH) !=
DJI_ERROR_SYSTEM_MODULE_CODE_SUCCESS) {
        printf("file system init error");
        return DJI_ERROR_SYSTEM_MODULE_CODE_UNKNOWN;
    }
    /* --------------------------------*/

    /* step 4 添加日志系统到控制台中-----------*/
    returnCode = DjiLogger_AddConsole(&printConsole);
    if (returnCode != DJI_ERROR_SYSTEM_MODULE_CODE_SUCCESS) {
        printf("add printf console error");
        return DJI_ERROR_SYSTEM_MODULE_CODE_SYSTEM_ERROR;
    }
    /* --------------------------------*/

    /* step 5 添加日志系统到日志文件中---------*/
    returnCode = DjiLogger_AddConsole(&localRecordConsole);
    if (returnCode != DJI_ERROR_SYSTEM_MODULE_CODE_SUCCESS) {
        printf("add printf console error");
        return DJI_ERROR_SYSTEM_MODULE_CODE_SYSTEM_ERROR;
    }
    /* --------------------- --------------*/
```

4. 修改 CMakeLists.txt

在 CMakeLists.txt 中需要修改 3 处，分别是添加电源管理的目录和文件、添加库文件。按照下面的提示找到相应的位置进行修改即可。

```
# 1 在此处加入电源管理目录下的所有.c 文件------------------------
file(GLOB_RECURSE MODULE_SAMPLE_SRC power_mgmt/*.c)

# 2 在此处加入电源管理目录------------------------------------
include_directories(./power_mgmt)

# 3 在此处加入树莓派 wiringPi 库文件/usr/lib/libwiringPi.so-----
```

```
link_libraries(/usr/lib/libwiringPi.so)
# --------------------------------------------------------
```
修改后保存并退出。

5. 修改 main.c

1）添加头文件

添加电源管理、日志管理和树莓派 GPIO 的头文件。

```c
#include <power_mgmt/power_mgmt.h>
#include <wiringPi.h>
#include <dji_logger.h>
```

2）设置树莓派

在 main 函数外部实现下面的功能。

```c
/* step 1 定义树莓派用于控制高电平的引脚，引脚为 3----------------------*/
static int powerPin = 3;                // 定义树莓派引脚3用于控制高电平
/* --------------------------------------------------------------*/

/* step 2 通过下面的函数完成树莓派 GPIO 引脚初始化，引脚为低电平输出模式--*/
static T_DjiReturnCode DjiTest_HighPowerApplyPinInit()
{
    if (wiringPiSetup() == -1)           // 初始化树莓派 GPIO，失败则返回-1
    {
        USER_LOG_ERROR("wiringPi GPIO setup error");
        return DJI_ERROR_SYSTEM_MODULE_CODE_UNKNOWN;
    }
    pinMode(powerPin, OUTPUT);           // 设置树莓派引脚为输出工作模式
    digitalWrite(powerPin, LOW);         // 初始化引脚为低电平

    return DJI_ERROR_SYSTEM_MODULE_CODE_SUCCESS;
}
/* --------------------------------------------------------------*/

/* step 3 通过下面的函数判断引脚是低电平还是高电平--------------------*/
static T_DjiReturnCode DjiTest_WriteHighPowerApplyPin
(E_DjiPowerManagementPinState pinState)
{
    switch (pinState)
    {
        case DJI_POWER_MANAGEMENT_PIN_STATE_RESET:
            digitalWrite(powerPin, LOW);      // 设置低电平
            break;
        case DJI_POWER_MANAGEMENT_PIN_STATE_SET:
            digitalWrite(powerPin, HIGH);     // 设置高电平
            break;
        default:
```

```
            USER_LOG_ERROR("pin state unknown: %d", pinState);
            return DJI_ERROR_SYSTEM_MODULE_CODE_UNKNOWN;
    }

    return DJI_ERROR_SYSTEM_MODULE_CODE_SUCCESS;
}
/* -------------------------------------------------------------- */
```

3）修改 main 函数

在 main 函数内部实现下面的功能。

```
int main(int argc, char **argv)
{

    /* --------在此处完成高功率申请的注册和电源管理功能的初始化------ */

    /* step 1 给 power_mgmt.h 中定义的结构体 T_DjiTestApplyHighPowerHandler 赋值-*/
    T_DjiTestApplyHighPowerHandler applyHighPowerHandler = {
        .pinInit = DjiTest_HighPowerApplyPinInit,
        .pinWrite = DjiTest_WriteHighPowerApplyPin,
    };
    /* -------------------------------------------------------------- */

    /* step 2 调用 DjiTest_RegApplyHighPowerHandler 函数注册申请高功率--*/
    returnCode = DjiTest_RegApplyHighPowerHandler(&applyHighPowerHandler);
    if (returnCode != DJI_ERROR_SYSTEM_MODULE_CODE_SUCCESS) {
        USER_LOG_ERROR("regsiter apply high power handler error");
    }
    /* -------------------------------------------------------------- */

    /* step 3 调用 DjiTest_PowerManagementStartService 函数启动电源管理服务----*/
    returnCode = DjiTest_PowerManagementStartService();
    if (returnCode != DJI_ERROR_SYSTEM_MODULE_CODE_SUCCESS) {
        USER_LOG_ERROR("power management init error");
    }
    /* -------------------------------------------------------------- */

}
```

6. 修改 power_mgmt.h

电源管理头文件无须修改。

7. 修改 power_mgmt.c

根据模板文件，在 DjiTest_PowerManagementStartService 函数下添加相应的代码。完成后的代码如下。

```
T_DjiReturnCode DjiTest_PowerManagementStartService(void)
{
```

```c
    T_DjiReturnCode returnCode;
    T_DjiAircraftInfoBaseInfo baseInfo = {0};


    /* step 1 调用 DjiPowerManagement_Init 函数完成电源管理模块的初始化----------- */
    returnCode = DjiPowerManagement_Init();
    if (returnCode != DJI_ERROR_SYSTEM_MODULE_CODE_SUCCESS) {
        USER_LOG_ERROR("power management init error: 0x%08llX.", returnCode);
        return returnCode;
    }
    /* ----------------------------------------------------------------------*/

    /* step 2 调用 DjiAircraftInfo_GetBaseInfo 函数获取无人机的参数--------------- */
    returnCode = DjiAircraftInfo_GetBaseInfo(&baseInfo);
    if (returnCode != DJI_ERROR_SYSTEM_MODULE_CODE_SUCCESS) {
        USER_LOG_ERROR("Get aircraft base info error: 0x%08llX.", returnCode);
        return returnCode;
    }
    /* -----------------------------------------------------------------------*/


    /* step 3 完成负载设备申请高功率 -------------------------------------*/
    if (baseInfo.aircraftType == DJI_AIRCRAFT_TYPE_M300_RTK &&
        (baseInfo.djiAdapterType == DJI_SDK_ADAPTER_TYPE_SKYPORT_V2 ||
         baseInfo.djiAdapterType == DJI_SDK_ADAPTER_TYPE_XPORT)) {

        /* step 3.1 获取树莓派初始化状态---------------*/
        s_applyHighPowerHandler.pinInit
        if (s_applyHighPowerHandler.pinInit == NULL) {
            USER_LOG_ERROR("apply high power pin init interface is NULL error");
            return DJI_ERROR_SYSTEM_MODULE_CODE_UNKNOWN;
        }
        /* ---------------------------------------------------*/


        /* step 3.2 获取当前高电平写入状态---------------------*/
        s_applyHighPowerHandler.pinWrite
        if (s_applyHighPowerHandler.pinWrite == NULL) {
            USER_LOG_ERROR("apply high power pin write interface is NULL error");
            return DJI_ERROR_SYSTEM_MODULE_CODE_UNKNOWN;
        }
        /* ---------------------------------------------------*/

        /* step 3.3 在此处初始化树莓派 ------------------------*/
        returnCode = s_applyHighPowerHandler.pinInit();
        if (returnCode != DJI_ERROR_SYSTEM_MODULE_CODE_SUCCESS) {
            USER_LOG_ERROR("apply high power pin init error");
            return returnCode;
```

            }
        /*--*/

        /* step 3.4 在此处通过 DjiPowerManagement_RegWriteHighPowerApplyPinCallback
回调函数设置高功率申请 -------*/
            returnCode = DjiPowerManagement_RegWriteHighPowerApplyPinCallback
(s_applyHighPowerHandler.pinWrite);
            if (returnCode != DJI_ERROR_SYSTEM_MODULE_CODE_SUCCESS) {
                USER_LOG_ERROR("register WriteHighPowerApplyPinCallback error.");
                return returnCode;
            }
        /* ---*/

        /* step 3.5 在此处通过调用 DjiPowerManagement_ApplyHighPowerSync 函数实现负载
设备能够向无人机申请高功率的功能 --*/
            returnCode = DjiPowerManagement_ApplyHighPowerSync();
            if (returnCode != DJI_ERROR_SYSTEM_MODULE_CODE_SUCCESS) {
                USER_LOG_ERROR("apply high power error");
                return returnCode;
            }
    }
        /* ---*/

    /* step 4 注册关机通知的 DjiPowerManagement_RegPowerOffNotificationCallback
回调函数---------------------*/
        returnCode = DjiPowerManagement_RegPowerOffNotificationCallback
(DjiTest_PowerOffNotificationCallback);
        if (returnCode != DJI_ERROR_SYSTEM_MODULE_CODE_SUCCESS) {
            USER_LOG_ERROR("register power off notification callback function error");
            return returnCode;
        }
     /* ---*/
        return DJI_ERROR_SYSTEM_MODULE_CODE_SUCCESS;
    }

    /* step 5 在此处实现关机通知的回调函数的关机功能--------------------------------*/
    static T_DjiReturnCode DjiTest_PowerOffNotificationCallback(bool
*powerOffPreparationFlag)
    {
        USER_LOG_INFO("aircraft will power off soon.");
        // 添加关机功能
        system("poweroff");
        *powerOffPreparationFlag = true;
      /* ---*/
    }

8. 编译并执行项目

```
# 创建 build 目录
pi@raspberrypi:~/pro_2_power_mgmt $ mkdir build && cd build

# 执行 cmake
pi@raspberrypi:~/pro_2_power_mgmt/build $ cmake ..

# 执行 make
pi@raspberrypi:~/pro_2_power_mgmt/build $ make

# 进入 build/bin 目录，执行程序
pi@raspberrypi:~/pro_2_power_mgmt/build $ cd bin/
pi@raspberrypi:~/pro_2_power_mgmt/build/bin $ sudo ./pro_2_power_mgmt
```

任务评价

任务过程评价表

任务实施人姓名_____ 学号_____ 时间_____

	评价项目及标准	分值	小组评议	教师评议
技术能力	1. 对基本概念的熟悉程度	10		
	2. 掌握日志管理功能的 PSDK 接口	10		
	3. 掌握电源管理功能的 PSDK 接口	10		
	4. 掌握控制台输出和日志输出的功能	10		
	5. 掌握负载设备高功率申请的功能	10		
	6. 掌握无人机对负载设备发送关机通知的方法	10		
执行能力	1. 出勤情况	5		
	2. 遵守纪律情况	5		
	3. 是否主动参与，有无提问记录	5		
	4. 有无职业意识	5		
社会能力	1. 能否有效沟通	5		
	2. 是否使用基本的文明礼貌用语	5		
	3. 能否与组员主动交流、积极合作	5		
	4. 能否自我学习、自我管理	5		
总分		100		
评定等级：				
评价意见		学习意见		

评定等级具体如下。

A：优，得分>90。B：好，90≥得分>80。C：一般，80≥得分>60。D：有待提高，得分≤60。

小结

本项目首先介绍了日志管理功能及其接口的使用方法，接着介绍了电源管理的两大功能（高功率申请和关机通知）及其接口的使用方法，最后介绍了相关代码模板。通过电源管理功能实现负载设备的高功率申请和关机通知功能，同时使用日志管理功能将相关的日志输出到控制台中。

项目 3

无人机信息管理

任务要求

本项目将使用 DJI PSDK API 进行消息订阅，负载设备可以获取订阅的无人机的速度、健康状态及 GPS 位置。

知识导入

3.1 信息管理功能

PSDK 信息管理功能包括信息获取功能和消息订阅功能，使用 PSDK 开发的负载设备具有信息获取功能，能够主动获取无人机的型号、负载设备挂载的位置、用户使用的移动端 App 等数据信息，加载不同的配置文件，方便用户使用负载设备；而具有消息订阅功能的负载设备能够记录用户订阅的数据信息，方便用户实现更广泛的应用。

1. 信息获取功能

信息获取功能是指负载设备能够主动获取并记录无人机的型号、硬件平台的类型和负载设备挂载的位置等数据信息。根据无人机实际的飞行状况，无人机上的各部件会实时产生大量的数据信息，这些信息将被无人机推送给其他模块。

用户使用具有消息订阅功能的负载设备能够指定所需订阅的数据信息。

2. 消息订阅功能

消息订阅功能可用于订阅无人机的数据信息，包括基础信息、GPS 信息和 RTK 信息等数据类型。每种数据类型都包含了多个订阅项，无人机订阅项如表 3-1 所示。

表 3-1　无人机订阅项

数据类型	订阅项（Topic）	最大订阅频率（Hz）
基础信息	姿态四元数	200
	速度	200
	融合海拔高度	200
	相对高度	200
	融合位置	200
	飞行状态	50
	电池信息	50
GPS 信息	GPS 日期	5
	GPS 时间	5
	GPS 位置	5
	GPS 速度	5
	GPS 信息	5
	GPS 信号等级	50
RTK 信息	RTK 位置	5
	RTK 速度	5
	RTK 航向角	5
	RTK 位置属性	5
	RTK 航向角属性	5

订阅数据项后，负载设备即可获取订阅的信息。消息订阅流程如图 3-1 所示。从图中可以看到，无人机与使用 PSDK 开发的负载设备通过 PSDK 接口完成了订阅数据推送和订阅数据获取。

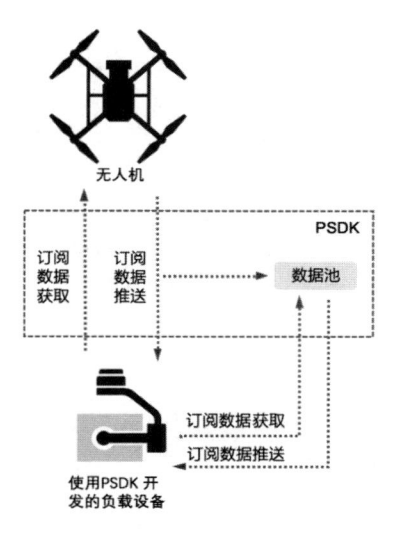

图 3-1　消息订阅流程

消息订阅功能最多支持订阅 4 类频率，如 1Hz、10Hz、20Hz、50Hz，且每个订阅项只能被订阅一次。订阅频率是指用户端接收数据的更新快慢，数值越大，获取的数据越灵敏，但不能超过该订阅项的最大订阅频率。在指定订阅频率时,任何参数的订阅频率都不能小于或等于 0Hz,相同订阅频率的主题的数据长度总和需小于或等于 242 字节。

3.2 接口说明

消息订阅相关功能的头文件为 dji_fc_subscription.h,该文件描述了消息订阅功能中结构体和函数原型的关键信息与使用方法。消息订阅需要使用下面 4 个函数原型。

- DjiFcSubscription_Init:消息订阅模块初始化。
- DjiFcSubscription_SubscribeTopic:以阻塞模式订阅消息。
- DjiFcSubscription_GetLatestValueOfTopic:获取最新的数据和时间戳。
- DjiFcSubscription_DeInit:释放消息订阅。

1. 模块初始化

在向无人机订阅任何数据信息之前,必须先初始化消息订阅模块,即 DjiFcSubscription_Init 函数必须先被调用,以初始化消息订阅模块。DjiFcSubscription_Init 函数定义如下。

```
T_DjiReturnCode DjiFcSubscription_Init();
```

不能在 main 函数中调用本函数,应在用户线程中调用本函数,启动调度器后,本函数将正常运行。根据程序的执行情况输出对应的返回值。返回的错误码定义在头文件 dji_error.h 中。

2. 订阅消息

PSDK 支持通过注册回调和接口调用两种方式订阅无人机对外推送的数据信息。

通过调用 DjiFcSubscription_SubscribeTopic 函数来指定订阅频率和订阅项;通过构造并注册回调函数,获取无人机最新产生的订阅项的数据信息及其对应的时间。该函数是以阻塞模式订阅消息的。DjiFcSubscription_SubscribeTopic 函数定义如下。

```
T_DjiReturnCode DjiFcSubscription_SubscribeTopic(
            E_DjiFcSubscriptionTopic topic,              // 订阅项的名称
            E_DjiDataSubscriptionTopicFreq frequency,    // 订阅频率
            DjiReceiveDataOfTopicCallback callback);     // 回调函数
```

说明:参数 topic 是订阅项的名称;参数 frequency 是订阅频率;参数 callback 用于接收订阅数据的回调函数。如果不需要回调函数,则此项可以设置为 NULL。

E_DjiDataSubscriptionTopicFreq 在头文件 dji_typedef.h 中被定义,具体如下。

```
typedef enum {
    DJI_DATA_SUBSCRIPTION_TOPIC_1_HZ = 1,
    DJI_DATA_SUBSCRIPTION_TOPIC_5_HZ = 5,
    DJI_DATA_SUBSCRIPTION_TOPIC_10_HZ = 10,
    DJI_DATA_SUBSCRIPTION_TOPIC_50_HZ = 50,
    DJI_DATA_SUBSCRIPTION_TOPIC_100_HZ = 100,
    DJI_DATA_SUBSCRIPTION_TOPIC_200_HZ = 200,
```

```
    DJI_DATA_SUBSCRIPTION_TOPIC_400_HZ = 400,
} E_DjiDataSubscriptionTopicFreq;
```

3. 获取消息

订阅信息后，可以通过 DjiFcSubscription_GetLatestValueOfTopic 函数获取无人机上订阅项最新的数据和时间戳。DjiFcSubscription_GetLatestValueOfTopic 函数定义如下。

```
T_DjiReturnCode DjiFcSubscription_GetLatestValueOfTopic(
E_DjiFcSubscriptionTopic topic,
            uint8_t *data, uint16_t dataSizeOfTopic,
            T_DjiDataTimestamp *timestamp);
```

说明：参数 topic 是订阅项的名称；参数 data 指向用于存储订阅项数据的存储空间，否则本函数将返回错误码；在正常情况下，参数 dataSizeOfTopic 的长度与订阅项的长度相同，否则可能导致内存溢出等问题；参数 timestamp 是获取的时间戳。如果无须获取时间戳，则该参数可为空。

3.3 代码模板

在开发者实现消息订阅功能时，不需要在 main 函数中调用消息订阅的接口，而需要创建用户线程来实现消息订阅功能。在用户线程中调用消息订阅的接口，启动调度器后，消息订阅接口将正常运行。本项目的代码模板位于 pro_3_source_template 目录中。

1. 代码结构

下面是实现消息订阅功能的代码框架，用户只需要编写用于实现消息订阅功能的 info_mgmt.c 和 info_mgmt.h 即可。

```
├── 3rdparty              // 第三方库文件目录，省略了该目录下的内容
├── CMakeLists.txt        // cmake 构建项目的文件
├── dji_init.c            // 系统初始化源文件
├── dji_init.h            // 系统初始化头文件
├── info_mgmt             // 用户实现消息订阅的目录
│   ├── info_mgmt.c       // 实现消息订阅的源文件
│   └── info_mgmt.h       // 头文件
├── hal                   // hal 文件目录，省略了该目录下的内容
├── main.c                // 主函数文件
├── monitor               // 监控文件目录，省略了该目录下的内容
├── osal                  // osal 文件目录，省略了该目录下的内容
└── psdk_lib              // psdk 库文件目录，省略了该目录下的内容
```

2. 头文件模板

本项目中的头文件名为 info_mgmt.h。该文件定义的内容可满足正常的应用开发需求，

如果需要实现特殊的功能，则可以按需进行修改。

头文件中定义了一个 DjiTest_FcSubscriptionStartService 函数。本项目中 info_mgmt.h 的模板文件无须修改。

3. 源文件模板

本项目的源文件名为 info_mgmt.c，该文件共定义了两个函数，下面的函数在 info_mgmt.h 中已被声明。

```
T_DjiReturnCode DjiTest_FcSubscriptionStartService(void)
```

另外一个函数是用户线程。

```
/* ---------------用户线程--------------------------------*/
static void *UserFcSubscription_Task(void *arg);
```

PSDK 支持通过注册回调和函数调用两种方式订阅无人机对外推送的数据信息。一种是通过调用 DjiFcSubscription_GetLatestValueOfTopic 函数直接获取无人机上订阅项最新的数据和时间戳；另一种是通过调用 DjiFcSubscription_SubscribeTopic 函数指定订阅频率和订阅项，通过构造并注册回调函数，以获取无人机最新产生的订阅项的数据信息及其对应时间。在默认情况下，DjiFcSubscription_SubscribeTopic 函数不使用回调函数。

本项目先通过调用 DjiFcSubscription_SubscribeTopic 函数指定订阅频率和订阅项，再通过调用 DjiFcSubscription_GetLatestValueOfTopic 函数获取无人机上订阅项最新的数据和时间戳。由于不以使用回调函数的方式获取订阅消息，因此本项目未使用回调函数。

本项目要重点理解 info_mgmt.c 中的两个函数。

- DjiTest_FcSubscriptionStartService 函数的功能是初始化消息订阅模块、订阅数据项，以及创建用户线程。
- 用户线程需要通过编写 UserFcSubscription_Task 函数来实现。

以下是 info_mgmt.c 的模板文件，按照提示即可完成代码的编写。

```
#include <utils/util_misc.h>
#include <math.h>
#include "dji_logger.h"
#include "dji_platform.h"

/*----------在此处加入本项目的头文件 info_mgmt.h--------------------*/


/*-----------------------------------------------------------------*/
#define FC_SUBSCRIPTION_TASK_FREQ              (1)
#define FC_SUBSCRIPTION_TASK_STACK_SIZE        (1024)

/* ---------用户线程----------------------------------------------*/
static void *UserFcSubscription_Task(void *arg);


/* ---------控制变量----------------------------------------------/
static T_DjiTaskHandle s_userFcSubscriptionThread;
static bool s_userFcSubscriptionVelocityShow = true;    // 飞行速度信息开关
static bool s_userFcSubscriptionGpsPositionShow = true; // GPS 位置信息开关
```

```c
static bool s_userFcSubscriptionGpsNumShow = true;     // GPS 数量信息开关
static bool s_userFcSubscriptionOtherDataShow = true;  // 其他信息开关

/* --------按照要求完成消息订阅模块初始化和消息订阅----------------------------*/

T_DjiReturnCode DjiTest_FcSubscriptionStartService(void)
{
    T_DjiReturnCode djiStat;
    T_DjiOsalHandler *osalHandler = NULL;

    osalHandler = DjiPlatform_GetOsalHandler();

    /*------调用 DjiFcSubscription_Init 函数初始化消息订阅模块----------*/
    /*------调用 DjiFcSubscription_SubscribeTopic 函数订阅速度数据项------*/
    /*------调用 DjiFcSubscription_SubscribeTopic 函数订阅 GPS 位置数据项--*/
    /*------调用 TaskCreate 函数创建用户线程,用于获取数据并将其输出到控制台中----------*/

    return DJI_ERROR_SYSTEM_MODULE_CODE_SUCCESS;
}

/* --------在用户线程中按照要求完成数据的获取与输出--------------------*/
static void *UserFcSubscription_Task(void *arg)
{
    T_DjiReturnCode djiStat;
    T_DjiFcSubscriptionVelocity velocity = {0};
    T_DjiDataTimestamp timestamp = {0};
    T_DjiFcSubscriptionGpsPosition gpsPosition = {0};
    T_DjiOsalHandler *osalHandler = NULL;

    USER_UTIL_UNUSED(arg);
    osalHandler = DjiPlatform_GetOsalHandler();

    while (1) {
        osalHandler->TaskSleepMs(1000 / FC_SUBSCRIPTION_TASK_FREQ);

        /*--此处调用 DjiFcSubscription_GetLatestValueOfTopic 函数获取速度数据---*/

        if (s_userFcSubscriptionVelocityShow == true)
        {
            /*--此处使用 USER_LOG_INFO 函数将速度数据输出到控制台中----------------*/

        }

        /*--此处调用 DjiFcSubscription_GetLatestValueOfTopic 函数获取 GPS 位置数据-*/

        if (s_userFcSubscriptionGpsPositionShow == true)
```

```
        {
            /*--此处使用 USER_LOG_INFO 函数将 GPS 位置数据输出到控制台中----------------*/
        }
    }
}
```

➡ 任务设计

速度订阅项是指地球中心坐标系（ground-fixed NEU frame）下的无人机速度，无人机速度的结构体参考了 T_DjiFcSubscriptionVelocity 数据结构，在该结构中定义了订阅项的健康状态。GPS 位置的宏定义参考了 T_DjiFcSubscriptionGpsPosition。

任务的具体设计要求如下。

（1）创建项目目录。

（2）复制 psdk_lib、hal、osal、monitor、3rdparty、main.c 和 CMakeLists.txt 到该项目目录下。

（3）本项目将实现消息订阅功能，项目目录下的 info_mgmt 目录用于编写代码，以实现用户线程调用消息订阅 API 接口。

（4）在 info_mgmt 目录下编写代码，实现消息订阅。

① 初始化消息订阅功能。

② 订阅数据项。

③ 获取订阅的数据项并将其输出到控制台中。

（5）使用 cmake 编译项目。

（6）执行项目。

➡ 任务实施

基于 PSDK 实现订阅和获取无人机的速度、健康状态和 GPS 位置，并将其输出到控制台中。限于篇幅，这里只给出任务实施步骤和关键代码。

1. 模板文件

连接树莓派开发板，默认路径为当前登录用户的家目录。查看当前的家目录，该目录下有一个名为 pro_3_source_template 的目录，此目录下保存了本项目的所有模板文件。

目录结构如下。

```
pi@raspberrypi:~ $ cd pro_3_source_template/
pi@raspberrypi:~/pro_3_source_template $ tree
.
├── 3rdparty              # 为目录，省略了目录下的文件
├── CMakeLists.txt        # cmake 构建项目文件
├── dji_init.c            # 系统初始化源文件
├── dji_init.h            # 系统初始化头文件
├── hal                   # 为目录，省略了目录下的文件
```

```
    ├── info_mgmt            # 消息订阅的目录
    │   ├── info_mgmt.c      # 消息订阅的源代码
    │   └── info_mgmt.h      # 消息订阅的头代码
    ├── main.c               # 主函数文件
    ├── monitor              # 为目录,省略了目录下的文件
    ├── osal                 # 为目录,省略了目录下的文件
    └── psdk_lib             # 为目录,省略了目录下的文件
```

2. 复制模板目录

将 pro_3_source_template 目录下的所有文件和目录复制到 pro_3_info_mgmt 目录中。

```
pi@raspberrypi:~ $ cp -r pro_3_source_template pro_3_info_mgmt
```

3. 修改 dji_init.c

在 dji_init.c 中填写用户账号信息,开发者需根据实际的 PSDK 用户参数进行填写。

```
/* ------按照下面的格式填写 PSDK 用户参数------------------------*/
#define USER_App_NAME              "RaspberryPi"
#define USER_App_ID                "126413"
#define USER_App_KEY               "0ea2eadb314d1a6e870975364e2bf4a"
#define USER_App_LICENSE           "HZ/GptycWQ………………"
#define USER_DEVELOPER_ACCOUNT     "abc@outlo**.com"
#define USER_BAUD_RATE             "460800"
```

编辑完成后保存并退出。

4. 修改 CMakeLists.txt

在 CMakeLists.txt 中需要修改 3 处,分别是修改项目的工程名、添加消息订阅的目录和文件。按照下文提示找到相应的位置进行修改即可。

```
pi@raspberrypi:~/pro_3_info_mgmt $ vim CMakeLists.txt

# --------设置工程名,将 my_project_name 修改为实际工程名----------------
project(pro_3_info_mgmt C)

# ---修改此处,将在项目下创建的消息订阅的目录及该目录下所有的 .c 文件加入项目-----
file(GLOB_RECURSE MODULE_SAMPLE_SRC info_mgmt/*.c)

# -------修改此处,将在项目下创建的消息订阅的目录加入项目-------------------
include_directories(./info_mgmt)
# -----------------------------------------------------------------
```

修改后保存并退出。

5. 修改 main.c

在 main.c 中需要修改两处,分别是添加消息订阅的头文件和初始化消息订阅模块。在 main 函数中添加消息订阅的头文件:

```
/*---------在此处添加消息订阅的头文件---------------------------*/
#include <info_mgmt/info_mgmt.h>
/* -----------------------------------------------------------*/
```

在 main 函数中添加初始化消息订阅功能的代码：

```
/*----------在此处添加初始化消息订阅功能的代码----------------------*/
returnCode = DjiTest_FcSubscriptionStartService();
if (returnCode != DJI_ERROR_SYSTEM_MODULE_CODE_SUCCESS) {
    USER_LOG_ERROR("data subscription init error\n");
}
/* ----------------------------------------------------------*/
```

修改后保存并退出。

6. 修改 info_mgmt.h

消息订阅的头文件无须修改。

7. 修改 info_mgmt.c

根据模板文件，添加相应的代码，完成后的代码如下。

```
/*----------在此处加入本项目的头文件----------------------------------*/
#include "info_mgmt.h"
/* ----------------------------------------------------------------*/

/* --------按照要求完成消息订阅功能初始化和消息订阅-------------------*/

T_DjiReturnCode DjiTest_FcSubscriptionStartService(void)
{
    T_DjiReturnCode djiStat;
    T_DjiOsalHandler *osalHandler = NULL;

    osalHandler = DjiPlatform_GetOsalHandler();

    /*-----------此处调用 DjiFcSubscription_Init 函数初始化消息订阅模块-*/
    djiStat = DjiFcSubscription_Init();
    if (djiStat != DJI_ERROR_SYSTEM_MODULE_CODE_SUCCESS) {
        USER_LOG_ERROR("init data subscription module error.");
        return DJI_ERROR_SYSTEM_MODULE_CODE_UNKNOWN;
    }
    /* ----------------------------------------------------------*/

    /*-----此处调用 DjiFcSubscription_SubscribeTopic 函数订阅速度数据项--*/
    djiStat = DjiFcSubscription_SubscribeTopic(
    DJI_FC_SUBSCRIPTION_TOPIC_VELOCITY,
    DJI_DATA_SUBSCRIPTION_TOPIC_1_HZ,
                                                NULL);
    if (djiStat != DJI_ERROR_SYSTEM_MODULE_CODE_SUCCESS) {
        USER_LOG_ERROR("Subscribe topic velocity error.");
        return DJI_ERROR_SYSTEM_MODULE_CODE_UNKNOWN;
    } else {
        USER_LOG_DEBUG("Subscribe topic velocity success.");
```

}
    /* -- */

    /*--此处调用 DjiFcSubscription_SubscribeTopic 函数订阅 GPS 位置数据项-*/
    djiStat = DjiFcSubscription_SubscribeTopic(
DJI_FC_SUBSCRIPTION_TOPIC_GPS_POSITION,
DJI_DATA_SUBSCRIPTION_TOPIC_1_HZ,
                                                NULL);
    if (djiStat != DJI_ERROR_SYSTEM_MODULE_CODE_SUCCESS) {
        USER_LOG_ERROR("Subscribe topic gps position error.");
        return DJI_ERROR_SYSTEM_MODULE_CODE_UNKNOWN;
    } else {
        USER_LOG_DEBUG("Subscribe topic gps position success.");
    }
    /* -- */

    /*----此处调用 TaskCreate 函数创建用户线程，用于获取数据并将其输出到控制台中---------*/
    if (osalHandler->TaskCreate("user_subscription_task",
UserFcSubscription_Task,
                                FC_SUBSCRIPTION_TASK_STACK_SIZE,
 NULL,
&s_userFcSubscriptionThread) !=
        DJI_ERROR_SYSTEM_MODULE_CODE_SUCCESS) {
        USER_LOG_ERROR("user data subscription task create error.");
        return DJI_ERROR_SYSTEM_MODULE_CODE_UNKNOWN;
    }
    /* -- */

    return DJI_ERROR_SYSTEM_MODULE_CODE_SUCCESS;
}

/* ---------在用户线程中按照要求完成数据获取和输出--------------------*/
static void *UserFcSubscription_Task(void *arg)
{
    T_DjiReturnCode djiStat;
    T_DjiFcSubscriptionVelocity velocity = {0};
    T_DjiDataTimestamp timestamp = {0};
    T_DjiFcSubscriptionGpsPosition gpsPosition = {0};
    T_DjiOsalHandler *osalHandler = NULL;

    USER_UTIL_UNUSED(arg);
    osalHandler = DjiPlatform_GetOsalHandler();

    while (1) {

```
            osalHandler->TaskSleepMs(1000 / FC_SUBSCRIPTION_TASK_FREQ);

            /*此处调用DjiFcSubscription_GetLatestValueOfTopic函数获取速度数据*/
            djiStat = DjiFcSubscription_GetLatestValueOfTopic(
    DJI_FC_SUBSCRIPTION_TOPIC_VELOCITY,
                                            (uint8_t *) &velocity,
                                sizeof(T_DjiFcSubscriptionVelocity),
                                                            timestamp);
            if (djiStat != DJI_ERROR_SYSTEM_MODULE_CODE_SUCCESS) {
                USER_LOG_ERROR("get value of topic velocity error.");
            }
            /* ---------------------------------------------------------*/

            if (s_userFcSubscriptionVelocityShow == true)
            {
                /*----此处使用USER_LOG_INFO函数将速度数据输出到控制台中------*/
                USER_LOG_INFO("velocity: x %f y %f z %f, healthFlag %d.",
velocity.data.x, velocity.data.y,velocity.data.z, velocity.health);
            }
            /* ---------------------------------------------------*/


            /*-此处调用DjiFcSubscription_GetLatestValueOfTopic函数获取GPS位置数据-*/
            djiStat = DjiFcSubscription_GetLatestValueOfTopic(
    DJI_FC_SUBSCRIPTION_TOPIC_GPS_POSITION,
                                            (uint8_t *) &gpsPosition,
                                sizeof(T_DjiFcSubscriptionGpsPosition),
                                                            &timestamp);
            if (djiStat != DJI_ERROR_SYSTEM_MODULE_CODE_SUCCESS) {
                USER_LOG_ERROR("get value of topic gps position error.");
            }

            if (s_userFcSubscriptionGpsPositionShow == true)
            {
                /*---此处使用USER_LOG_INFO函数将GPS位置数据输出到控制台中----*/
                USER_LOG_INFO("gps position: x %d y %d z %d.", gpsPosition.x,
gpsPosition.y, gpsPosition.z);
                /* ---------------------------------------------------*/

            }
        }
    }
```

8. 编译并执行项目

```
# 在项目pro_3_info_mgmt目录下，创建build目录，并进入build目录
pi@raspberrypi:~/pro_3_info_mgmt $ mkdir build && cd build
```

```
# 使用 cmake 创建项目编译环境
pi@raspberrypi:~/pro_3_info_mgmt/build $ cmake ..

# make 编译
pi@raspberrypi:~/pro_3_info_mgmt/build $ make

# 执行文件位于 build/bin 目录下
pi@raspberrypi:~/pro_3_info_mgmt/build $ cd bin/

# 使用 sudo 执行项目的执行文件
pi@raspberrypi:~/pro_3_info_mgmt/build/bin $ sudo ./pro_3_info_mgmt
```

任务评价

任务过程评价表

任务实施人姓名_____ 学号_____ 时间_____

	评价项目及标准	分值	小组评议	教师评议
技术能力	1. 对基本概念的熟悉程度	10		
	2. 掌握消息订阅 PSDK 接口	10		
	3. 掌握消息订阅的代码框架	10		
	4. 掌握消息订阅模块初始化功能	10		
	5. 掌握消息订阅用户线程的使用方法	10		
	6. 掌握基于代码模板的修改与添加方法	10		
执行能力	1. 出勤情况	5		
	2. 遵守纪律情况	5		
	3. 是否主动参与，有无提问记录	5		
	4. 有无职业意识	5		
社会能力	1. 能否有效沟通	5		
	2. 是否使用基本的文明礼貌用语	5		
	3. 能否与组员主动交流、积极合作	5		
	4. 能否自我学习、自我管理	5		
总分		100		
评定等级：				
评价意见		学习意见		

评定等级具体如下。

A：优，得分>90。B：好，90≥得分>80。C：一般，80≥得分>60。D：有待提高，得分≤60。

小结

本项目首先简要介绍了 DJI 无人机消息订阅的相关知识及可订阅的数据项；其次介绍了与消息订阅相关的接口；最后介绍了消息订阅的代码框架。

项目 4 相机管理与云台管理

任务 1 相机管理

任务要求

相机管理的接口能够同时设置并获取无人机上多个相机的感光度、光圈、快门和曝光等参数值，控制相机实现拍照、录像及指点变焦等功能。本任务基于相机管理的接口控制负载相机进行拍照和录像。

知识导入

4.1 相机管理功能

相机管理（camera manager）是 OSDK 端口功能，与 PSDK 挂载的相机（camera emu）不一样。PSDK 挂载的相机用于集成一个挂载到无人机上的第三方相机负载，而 OSDK 端口的相机管理用于对负载端口的相机进行控制。相机管理可以控制 M300 适配的 DJI 相机，支持通过 PSDK 挂载的相机接入 M300 的相机。

相机管理的主要功能有拍照、录像、指定变焦等，以及通过 API 访问并下载相机 SD 卡中的照片等文件。

4.2 接口说明

相机管理相关功能的头文件为 camera_manager.h，该文件描述了相机管理功能中结构体和函数原型的关键信息与使用方法。相机管理功能函数如表 4-1 所示。

表 4-1 相机管理功能函数

功能函数	功能
DjiCameraManager_Init	相机管理模块初始化
DjiCameraManager_DeInit	相机管理模块去初始化
DjiCameraManager_GetCameraType	获取所选位置相机的摄像机类型
DjiCameraManager_GetFirmwareVersion	获取所选位置相机的固件版本
DjiCameraManager_SetMode	设置所选位置相机的工作模式
DjiCameraManager_GetMode	获取所选位置相机的工作模式
DjiCameraManager_SetShootPhotoMode	设置所选位置相机的拍照模式
DjiCameraManager_GetShootPhotoMode	获取所选位置相机的拍照模式
DjiCameraManager_StartShootPhoto	开始拍照
DjiCameraManager_StopShootPhoto	停止拍照
DjiCameraManager_SetPhotoBurstCount	设置在连拍模式下的连拍次数
DjiCameraManager_GetPhotoBurstCount	获取在连拍模式下的连拍次数
DjiCameraManager_SetPhotoAEBCount	设置在 AEB（自动包围曝光）拍照模式中的连拍次数
DjiCameraManager_GetPhotoAEBCount	获取在 AEB（自动包围曝光）拍照模式中的连拍次数
DjiCameraManager_SetPhotoTimeIntervalSettings	设置在间隔拍照模式下的连拍次数
DjiCameraManager_GetPhotoTimeIntervalSettings	获取在间隔拍照模式下的连拍次数
DjiCameraManager_SetFocusMode	设置所选位置相机的对焦模式
DjiCameraManager_GetFocusMode	获取所选位置相机的对焦模式
DjiCameraManager_SetFocusTarget	设置所选位置相机的焦点
DjiCameraManager_GetFocusTarget	获取所选位置相机的焦点
DjiCameraManager_StartContinuousOpticalZoom	启动所选位置相机的光学变焦
DjiCameraManager_StopContinuousOpticalZoom	停止所选位置相机的光学变焦
DjiCameraManager_SetOpticalZoomParam	设置所选位置相机的摄像机光学变焦参数
DjiCameraManager_GetOpticalZoomParam	获取所选位置相机的摄像机光学变焦参数
DjiCameraManager_SetTapZoomEnabled	启用/禁用所选位置相机的摄像机指点变焦功能
DjiCameraManager_GetTapZoomEnabled	获取所选位置相机的摄像机指点变焦功能的状态
DjiCameraManager_SetTapZoomMultiplier	设置所选位置相机的摄像机指点变焦倍数
DjiCameraManager_GetTapZoomMultiplier	获取所选位置相机的摄像机指点变焦倍数
DjiCameraManager_TapZoomAtTarget	设置所选位置相机的摄像机指点变焦点
DjiCameraManager_SetExposureMode	设置所选位置相机的曝光模式

续表

功能函数	功能
DjiCameraManager_GetExposureMode	获取所选位置相机的曝光模式
DjiCameraManager_SetISO	设置所选位置相机的 ISO 数值
DjiCameraManager_GetISO	获取所选位置相机的 ISO 数值
DjiCameraManager_SetAperture	设置所选位置相机的光圈大小
DjiCameraManager_GetAperture	获取所选位置相机的光圈大小
DjiCameraManager_SetShutterSpeed	设置所选位置相机的快门速度
DjiCameraManager_GetShutterSpeed	获取所选位置相机的快门速度
DjiCameraManager_SetExposureCompensation	设置所选位置相机的曝光补偿数值
DjiCameraManager_GetExposureCompensation	获取所选位置相机的曝光补偿数值

以下是部分函数原型的具体说明，更多说明请参阅大疆官网。

1. 相机管理模块初始化

```
T_DjiReturnCode DjiCameraManager_Init(void);
```

2. 相机管理模块去初始化

```
T_DjiReturnCode DjiCameraManager_DeInit(void);
```

3. 获取所选位置相机的摄像机类型

```
T_DjiReturnCode DjiCameraManager_GetCameraType(
            E_DjiMountPosition position,
            E_DjiCameraType *cameraType);
```

说明：参数 position 是所选相机的位置；参数 cameraType 的说明参见 E_DjiCameraType 的参考资料。

4. 获取所选位置相机的固件版本

```
T_DjiReturnCode DjiCameraManager_GetFirmwareVersion(
            E_DjiMountPosition position,
            T_DjiCameraManagerFirmwareVersion *firmwareVersion);
```

说明：参数 position 是所选相机的位置；参数 firmwareVersion 的说明参见 T_DjiCameraManagerFirmwareVersion 的参考资料。

5. 设置所选位置相机的工作模式

```
T_DjiReturnCode DjiCameraManager_SetMode(
            E_DjiMountPosition position,
            E_DjiCameraManagerWorkMode workMode);
```

说明：参数 position 是所选相机的位置；参数 workMode 的说明参见 E_DjiCameraManagerWorkMode 的参考资料。

4.3 代码模板

1. 代码结构

下面是实现相机管理功能的代码框架，用户只需要编写用于实现相机管理功能的 camera_manager.c 和 camera_manager.h 即可。

```
.
├── 3rdparty
├── camera_manager
│   ├── camera_manager.c
│   └── camera_manager.h
├── CMakeLists.txt
├── dji_init.c
├── dji_init.h
├── hal
├── main.c
├── monitor
├── osal
└── psdk_lib
```

2. 头文件模板

本任务的头文件名为 camera_manager.h。该文件定义的内容可满足正常的应用开发需求，无须修改。如果需要实现特殊的功能，则可以按需要修改。

3. 源文件模板

本任务的源文件名为 camera_manager.c，我们需要在 DjiTest_CameraManagerRunSample 相机管理服务入口函数中添加所有需要的功能代码。

```
/*----在相机管理服务入口函数中完成以下功能--------------*/
/* step 1 初始化相机管理功能----------------------*/
/* step 2 获取相机类型--------------------------*/
/* step 3 获取相机硬件版本-----------------------*/
/* step 4 实现单拍功能--------------------------*/
/* step 5 实现录像功能--------------------------*/

/*-------------------------------------------*/

T_DjiReturnCode DjiTest_CameraManagerRunSample(
E_DjiMountPosition mountPosition,
        E_DjiTestCameraManagerSampleSelect cameraManagerSampleSelect)
{
    T_DjiOsalHandler *osalHandler = DjiPlatform_GetOsalHandler();
```

```c
        T_DjiReturnCode returnCode;
        E_DjiCameraType cameraType;
        T_DjiCameraManagerFirmwareVersion firmwareVersion;
        T_DjiCameraManagerFocusPosData focusPosData;
        T_DjiCameraManagerTapZoomPosData tapZoomPosData;

        USER_LOG_INFO("Camera manager sample start");
    USER_LOG_INFO("--> Step 1: Init camera manager module");

        /* step 1 初始化相机管理功能----------------------*/

        /* ---------------------------------------------*/

        USER_LOG_INFO("--> Step 2: Get camera type and version");

        /* step 2 获取相机类型--------------------------*/

        /* ---------------------------------------------*/
        USER_LOG_INFO("Mounted position %d camera's type is %s",
                  mountPosition,
                  s_cameraTypeStrList[DjiTest_CameraManagerGetCameraTypeIndex(cameraType)].cameraTypeStr);

        /* step 3 获取相机硬件版本----------------------*/

        /* ---------------------------------------------*/
        USER_LOG_INFO("Mounted position %d camera's firmware is V%02d.%02d.%02d.%02d\r\n", mountPosition,
                  firmwareVersion.firmware_version[0],
firmwareVersion.firmware_version[1],
                  firmwareVersion.firmware_version[2],
firmwareVersion.firmware_version[3]);

        ---------------省略中间部分代码------------------

    case E_DJI_TEST_CAMERA_MANAGER_SAMPLE_SELECT_SHOOT_SINGLE_PHOTO: {
            USER_LOG_INFO("--> Function h: Shoot Single photo");
            /* step 4 实现单拍功能--------------------------*/

            /*---------------------------------------------*/
            break;
        }

    case E_DJI_TEST_CAMERA_MANAGER_SAMPLE_SELECT_RECORD_VIDEO: {
            USER_LOG_INFO("--> Function l: Record video in next 10s");
            /* step 5 实现录像功能--------------------------*/
```

```
        /*----------------------------------------------*/

        USER_LOG_INFO("Sleep 10s...");
        osalHandler->TaskSleepMs(10000);

        returnCode = DjiTest_CameraManagerStopRecordVideo(mountPosition);
        if (returnCode != DJI_ERROR_SYSTEM_MODULE_CODE_SUCCESS) {
            USER_LOG_ERROR("Mounted position %d camera stop record video failed, error code: 0x%08X\r\n",
                           mountPosition, returnCode);
            goto exitCameraModule;
        }
        break;
    }
---------------省略后面所有代码----------------------------------
```

任务设计

本任务基于相机管理的接口控制负载相机进行拍照和录像。任务的具体设计要求如下。

（1）从 pro_4_camera_mgmt_source_template 项目模板文件中复制一个新项目。文件名为 pro_4_camera_mgmt。

（2）本任务将实现相机管理功能，项目目录下的 camera_manager 目录用于编写代码，以实现相机管理功能 API 接口。

（3）在 camera_manager 目录下编写代码，实现相机管理功能。

① 初始化相机管理功能。
② 获取相机类型。
③ 获取相机硬件版本。
④ 实现单拍功能。
⑤ 实现录像功能。

（4）使用 cmake 编译项目。

（5）执行项目。

任务实施

限于篇幅，这里只给出任务实施步骤和关键代码。

1. 模板文件

连接树莓派开发板，默认路径为当前登录用户的家目录。查看当前的家目录，该目录下有一个名为 pro_4_camera_mgmt_source_template 的目录，此目录下保存了本任务的所有模板文件。

目录结构如下。

```
pi@raspberrypi:~ $ cd pro_4_camera_mgmt_source_template/
pi@raspberrypi:~/ pro_4_camera_mgmt_source_template $ tree
.
├── 3rdparty
├── camera_manager
│   ├── camera_manager.c
│   └── camera_manager.h
├── CMakeLists.txt
├── dji_init.c
├── dji_init.h
├── hal
├── main.c
├── monitor
├── osal
└── psdk_lib
```

2. 复制模板目录

将 pro_4_camera_mgmt_source_template 目录下的所有文件和目录复制到 pro_4_camera_mgmt 目录中。

```
pi@raspberrypi:~ $ cp -r pro_4_camera_mgmt_source_template pro_4_camera_mgmt
pi@raspberrypi:~ $ cd pro_4_camera_mgmt
```

3. 修改 dji_init.c

在 dji_init.c 中填写用户账号信息，开发者需根据实际的 PSDK 用户参数填写。

```
/* ------按照下面的格式填写 PSDK 用户参数----------------------------*/
#define USER_App_NAME              "RaspberryPi"
#define USER_App_ID                "126413"
#define USER_App_KEY               "0ea2eadb314d1a6e870975364e2bf4a"
#define USER_App_LICENSE           "HZ/GptycWQ........................"
#define USER_DEVELOPER_ACCOUNT     "abc@outlo**.com"
#define USER_BAUD_RATE             "460800"
```

4. 修改 CMakeLists.txt

（1）设置本项目的工程名为 pro_4_camera_mgmt。

```
# --------设置工程名，将 my_project_name 修改为实际工程名--------------
project(pro_4_camera_mgmt C)
# -----------------------------------------------------------
```

（2）添加目录、文件。

```
# -------在此处添加相机管理目录及该目录下的所有.c 文件-----------------
file(GLOB_RECURSE MODULE_SAMPLE_SRC camera_manager/*.c)
include_directories(./camera_manager)
# -----------------------------------------------------------
```

5. 修改 main.c

（1）添加头文件。

```
/*----------在此处添加相机管理功能的头文件-----------------------------*/
#include <camera_manager/camera_manager.h>
/*-----------------------------------------------------------------*/
```

（2）在 main 函数中设置相机位置、单拍和录像功能。

```
int main(int argc, char **argv)
{

    /*----在此处设置相机位置和相机管理服务入口函数---------------------*/
    static E_DjiMountPosition mountPosition = DJI_MOUNT_POSITION_PAYLOAD_PORT_NO1;

    /* step 1 在此处添加代码，设置单拍功能-----------------------------*/
    returnCode = DjiTest_CameraManagerRunSample(
        mountPosition,
        E_DJI_TEST_CAMERA_MANAGER_SAMPLE_SELECT_SHOOT_SINGLE_PHOTO);
    if (returnCode != DJI_ERROR_SYSTEM_MODULE_CODE_SUCCESS) {
        USER_LOG_ERROR("Camera Manager init error\n");
    }
    /* -------------------------------------------------------------*/

    /* step 2 在此处添加代码，设置录像功能-----------------------------*/
    returnCode = DjiTest_CameraManagerRunSample(
        mountPosition,
        E_DJI_TEST_CAMERA_MANAGER_SAMPLE_SELECT_RECORD_VIDEO);
    if (returnCode != DJI_ERROR_SYSTEM_MODULE_CODE_SUCCESS) {
        USER_LOG_ERROR("Camera Manager init error\n");
    }
    /* -------------------------------------------------------------*/
```

6. 修改 camera_manager.h

头文件无须修改。

7. 修改 camera_manager.c

根据 camera_manager.c 模板文件，需要完成以下功能。

```
/*----在相机管理服务入口函数中完成以下功能---------------*/

    /* step 1 初始化相机管理功能-------------------------*/
    returnCode = DjiCameraManager_Init();
    if (returnCode != DJI_ERROR_SYSTEM_MODULE_CODE_SUCCESS) {
        USER_LOG_ERROR("Init camera manager failed, error code: 0x%08X\r\n", returnCode);
        goto exitCameraModule;
    }
    /* ------------------------------------------------*/
```

```c
        /* step 2 获取相机类型--------------------------------*/
        returnCode = DjiCameraManager_GetCameraType(mountPosition, &cameraType);
        if (returnCode != DJI_ERROR_SYSTEM_MODULE_CODE_SUCCESS) {
            USER_LOG_ERROR("Get mounted position %d camera's type failed, error code: 0x%08X\r\n",
                           mountPosition, returnCode);
            goto exitCameraModule;
        }
        /* ----------------------------------------------*/

        /* step 3 获取相机硬件版本--------------------------*/
    returnCode = DjiCameraManager_GetFirmwareVersion(
mountPosition, &firmwareVersion);
        if (returnCode != DJI_ERROR_SYSTEM_MODULE_CODE_SUCCESS) {
            USER_LOG_ERROR("Get mounted position %d camera's firmware version failed, error code: 0x%08X\r\n",
                           mountPosition, returnCode);
            goto exitCameraModule;
        }
    /* ----------------------------------------------*/

    case E_DJI_TEST_CAMERA_MANAGER_SAMPLE_SELECT_SHOOT_SINGLE_PHOTO: {
            USER_LOG_INFO("--> Function h: Shoot Single photo");

            /* step 4 实现单拍功能---------------------------*/
            returnCode = DjiTest_CameraManagerStartShootSinglePhoto(
    mountPosition);
            if (returnCode != DJI_ERROR_SYSTEM_MODULE_CODE_SUCCESS) {
                USER_LOG_ERROR("Mounted position %d camera shoot single photo failed,error code: 0x%08X\r\n", mountPosition, returnCode);
                goto exitCameraModule;
            }
            break;
            /*----------------------------------------------*/
        }
    case E_DJI_TEST_CAMERA_MANAGER_SAMPLE_SELECT_RECORD_VIDEO: {
            USER_LOG_INFO("--> Function l: Record video in next 10s");

            /* step 5 实现录像功能---------------------------*/
            returnCode = DjiTest_CameraManagerStartRecordVideo(
    mountPosition);
            if (returnCode != DJI_ERROR_SYSTEM_MODULE_CODE_SUCCESS) {
                USER_LOG_ERROR("Mounted position %d camera start record video failed, error code: 0x%08X\r\n",mountPosition, returnCode);
                goto exitCameraModule;
```

```
        }
        /*----------------------------------------------*/
```

8. 编译并执行项目

```
# 在项目 pro_4_camera_mgmt 目录下，创建 build 目录，并进入 build 目录
pi@raspberrypi:~/pro_4_camera_mgmt $ mkdir build && cd build

# 使用 cmake 创建项目编译环境
pi@raspberrypi:~/pro_4_camera_mgmt/build $ cmake ..

# make 编译
pi@raspberrypi:~/pro_4_camera_mgmt/build $ make

# 执行文件位于 build/bin 目录下
pi@raspberrypi:~/pro_4_camera_mgmt/build $ cd bin/

# 使用 sudo 执行项目的执行文件
pi@raspberrypi:~/pro_4_camera_mgmt/build/bin $ sudo ./pro_4_camera_mgmt
```

任务评价

任务过程评价表

任务实施人姓名_____ 学号_____ 时间_____

	评价项目及标准	分值	小组评议	教师评议
技术能力	1. 对基本概念的熟悉程度	10		
	2. 掌握相机管理功能接口	10		
	3. 掌握相机管理功能的代码框架	10		
	4. 掌握相机管理模块初始化功能	10		
	5. 掌握相机管理功能中单拍的实现方法	10		
	6. 掌握相机管理功能中录像的实现方法	10		
执行能力	1. 出勤情况	5		
	2. 遵守纪律情况	5		
	3. 是否主动参与，有无提问记录	5		
	4. 有无职业意识	5		
社会能力	1. 能否有效沟通	5		
	2. 是否使用基本的文明礼貌用语	5		
	3. 能否与组员主动交流、积极合作	5		
	4. 能否自我学习、自我管理	5		
总分		100		
评定等级：				
评价意见		学习意见		

评定等级具体如下。

A：优，得分>90。B：好，90≥得分>80。C：一般，80≥得分>60。D：有待提高，得分≤60。

小结

本任务首先简要介绍了 DJI 无人机相机管理功能的相关知识；其次介绍了与相机管理功能相关的接口；最后介绍了相机管理功能的代码框架。

任务 2　云台管理

任务要求

本任务将基于云台管理，实现在无人机飞行的过程中控制指定的云台旋转对应角度，并回到中位。

知识导入

4.4　云台管理功能

云台管理在 M300 上为 OSDK 端口功能，主要用于机载计算机控制云台转动等控制功能，区别于 PSDK 端口的 gimbal emu（开发集成的第三方云台负载）。云台管理功能相对简单，在 OSDK 端口通过 API 的方式提供了云台转动的接口。

OSDK 端口功能可以控制挂载在 M300 上的 DJI 官方相机云台，如 H20/T；也可以控制基于 PSDK 开发集成的第三方负载云台。DJI 标准云台 X-PORT 可以通过 OSDK 端口的 gimbal manager 控制云台在 roll、pitch、yaw 方向上的旋转，包括回中。

云台的 3 种模式如下。

（1）自由模式：当无人机的姿态发生改变时，云台将不会转动。

（2）FPV 模式：当无人机的姿态发生改变时，云台会转动航向轴与横滚轴，确保负载设备当前的视场角不会发生改变。

（3）YAW 跟随模式：在该模式下，云台的航向轴会跟随无人机的航向轴转动。

云台控制模式如下。

（1）相对角模式：基于云台的当前角度，旋转指定的角度。

（2）绝对角模式：基于 NEU 坐标系，将云台旋转到指定的角度。

（3）速度模式：基于指定的速度旋转云台 0.5s。

4.5 接口说明

云台管理功能的头文件为 dji_gimbal_manager.h，该文件描述了云台管理功能中结构体和函数原型的关键信息与使用方法。云台管理功能定义了很多函数，如表 4-2 所示。

表 4-2　云台管理功能函数

函数	功能
DjiGimbalManager_Init	初始化云台管理模块
DjiGimbalManager_Deinit	去初始化云台管理模块
DjiGimbalManager_SetMode	设置云台的工作模式
DjiGimbalManager_Reset	云台的俯仰轴和偏航轴转动至中位
DjiGimbalManager_Rotate	重置云台的旋转

1. 初始化云台管理模块

```
T_DjiReturnCode DjiGimbalManager_Init(void);
```

2. 去初始化云台管理模块

```
T_DjiReturnCode DjiGimbalManager_Deinit(void)
```

3. 设置云台的工作模式

```
T_DjiReturnCode DjiGimbalManager_SetMode(
                    E_DjiMountPosition mountPosition,
                    E_DjiGimbalMode mode);
```

说明：参数 mountPosition 为云台安装位置，输入限制见枚举 E_DjiMountPosition；参数 mode 为云台工作模式，输入限制见枚举 E_DjiGimbalMode。

```
typedef enum {
    DJI_GIMBAL_MODE_FREE = 0,           // 自由模式
    DJI_GIMBAL_MODE_FPV = 1,            // FPV 模式
    DJI_GIMBAL_MODE_YAW_FOLLOW = 2,     // YAW 跟随模式
} E_DjiGimbalMode;
```

4. 云台的俯仰轴和偏航轴转动至中位

```
T_DjiReturnCode DjiGimbalManager_Reset(E_DjiMountPosition mountPosition);
```
说明：参数 mountPosition 为云台安装位置。

5. 重置云台的旋转

```
T_DjiReturnCode DjiGimbalManager_Rotate(
                    E_DjiMountPosition mountPosition,
                    T_DjiGimbalManagerRotation rotation);
```

说明：参数 mountPosition 为云台安装位置；参数 rotation 为目标云台要执行的旋转参数，包括旋转方式、目标角度值和执行时间等。

4.6 代码模板

1. 代码结构

下面是实现云台管理功能的代码框架，用户只需要编写用于实现云台管理功能的 gimbal_manager.c 和 gimbal_manager.h 即可。

```
.
├── 3rdparty
├── CMakeLists.txt
├── dji_init.c
├── dji_init.h
├── gimbal_manager
│   ├── gimbal_manager.c
│   └── gimbal_manager.h
├── hal
├── main.c
├── monitor
├── osal
└── psdk_lib
```

2. 头文件模板

本任务的头文件名为 gimbal_manager.h。该文件定义的内容可满足正常的应用开发需求，无须修改。如果需要实现特殊的功能，则可以按需进行修改。

3. 源文件模板

本任务的源文件名为 gimbal_manager.c，在 DjiTest_GimbalManagerRunSample 云台管理服务入口函数中添加所有需要的功能代码。

```
/*------在云台管理服务入口函数中完成以下功能------------*/
/* step 1 初始化云台管理功能----------------------*/
/* step 2 设置云台模式---------------------------*/
/* step 3 设置云台回中---------------------------*/
/* step 4 设置云台旋转---------------------------*/
/* ---------------------------------------------*/
T_DjiReturnCode DjiTest_GimbalManagerRunSample(E_DjiMountPosition mountPosition, E_DjiGimbalMode gimbalMode)
{
    T_DjiOsalHandler *osalHandler = DjiPlatform_GetOsalHandler();
    T_DjiReturnCode returnCode;
    T_DjiGimbalManagerRotation rotation;
```

```c
        USER_LOG_INFO("Gimbal manager sample start");
        USER_LOG_INFO("--> Step 1: Init gimbal manager module");
        /* step 1 初始化云台管理功能--------------------------*/

        /*------------------------------------------------*/
        if (gimbalMode == DJI_GIMBAL_MODE_FREE) {
            USER_LOG_INFO("--> Step 2: Set gimbal to free mode");
        } else if (gimbalMode == DJI_GIMBAL_MODE_YAW_FOLLOW) {
            USER_LOG_INFO("--> Step 2: Set gimbal to yaw follow mode");
        }
        /* step 2 设置云台模式------------------------------*/

        /*------------------------------------------------*/
        USER_LOG_INFO("--> Step 3: Reset gimbal angles.\r\n");
        /* step 3 设置云台回中------------------------------*/

        /*------------------------------------------------*/
        USER_LOG_INFO("--> Step 4: Rotate gimbal to target angle by action list\r\n");
        for (int i = 0; i < sizeof(s_rotationActionList) /
sizeof(T_DjiTestGimbalActionList); ++i) {
            if (s_rotationActionList[i].action == DJI_TEST_GIMBAL_RESET) {
                USER_LOG_INFO("Target gimbal reset.\r\n");
                returnCode = DjiGimbalManager_Reset(mountPosition);
                if (returnCode != DJI_ERROR_SYSTEM_MODULE_CODE_SUCCESS) {
                    USER_LOG_ERROR("Reset gimbal failed, error code: 0x%08X",
returnCode);
                }
                osalHandler->TaskSleepMs(2000);
            } else if (s_rotationActionList[i].action == DJI_TEST_GIMBAL_ROTATION) {

                if (gimbalMode == DJI_GIMBAL_MODE_FREE &&
                    s_rotationActionList[i].rotation.rotationMode ==
DJI_GIMBAL_ROTATION_MODE_ABSOLUTE_ANGLE) {
                    continue;
                }

                USER_LOG_INFO("Target gimbal pry = (%.1f, %.1f, %.1f)",
                              s_rotationActionList[i].rotation.pitch,
s_rotationActionList[i].rotation.roll,
                              s_rotationActionList[i].rotation.yaw);
                rotation = s_rotationActionList[i].rotation;

                /* step 4 设置云台旋转------------------------------*/

                /* ------------------------------------------------*/
                osalHandler->TaskSleepMs(1000);
            }
```

```
        }
        USER_LOG_INFO("--> Step 5: Deinit gimbal manager module");
        returnCode = DjiGimbalManager_Deinit();
        if (returnCode != DJI_ERROR_SYSTEM_MODULE_CODE_SUCCESS) {
            USER_LOG_ERROR("Deinit gimbal manager failed, error code: 0x%08X",
returnCode);
            goto out;
        }
    out:
        USER_LOG_INFO("Gimbal manager sample end");

        return returnCode;
    }
```

● 任务设计

本任务是在相机管理功能的基础上完成云台管理功能的。任务的具体设计要求如下。

（1）从 pro_4_gimbal_mgmt_source_template 项目模板文件中复制一个新项目。项目的工程名为 pro_4_gimbal_mgmt。

（2）本任务将实现云台管理功能，项目目录下的 gimbal_manager 目录用于编写代码，以实现云台管理功能 API 接口。

（3）在 gimbal_manager 目录下编写代码，实现云台管理功能。

① 初始化云台管理功能。
② 设置云台模式。
③ 设置云台回中。
④ 控制云台旋转。

（4）使用 cmake 编译项目。

（5）执行项目。

● 任务实施

限于篇幅，这里只给出任务实施步骤和关键代码。

1. 模板文件

连接树莓派开发板，默认路径为当前登录用户的家目录。查看当前的家目录，该目录下有一个名为 pro_4_gimbal_mgmt_source_template 的目录，此目录下保存了本任务的所有模板文件。

目录结构如下。

```
pi@raspberrypi:~ $ cd pro_4_gimbal_mgmt_source_template/
pi@raspberrypi:~/ pro_4_gimbal_mgmt_source_template $ tree
.
├── 3rdparty
├── CMakeLists.txt
```

```
├── dji_init.c
├── dji_init.h
├── gimbal_manager
│   ├── gimbal_manager.c
│   └── gimbal_manager.h
├── hal
├── main.c
├── monitor
├── osal
└── psdk_lib
```

2. 复制模板目录

将 pro_4_gimbal_mgmt_source_template 目录下的所有文件和目录复制到 pro_4_gimbal_mgmt 目录中。

```
pi@raspberrypi:~ $ cp -r pro_4_gimbal_mgmt_source_template pro_4_gimbal_mgmt
pi@raspberrypi:~ $ cd pro_4_gimbal_mgmt
```

3. 修改 dji_init.c

在 dji_init.c 中填写用户账号信息，开发者需根据实际的 PSDK 用户参数填写。

```
#define USER_App_NAME              "RaspberryPi"
#define USER_App_ID                "126413"
#define USER_App_KEY               "0ea2eadb314d1a6e870975364e2bf4a"
#define USER_App_LICENSE           "HZ/GptycWQ........................"
#define USER_DEVELOPER_ACCOUNT     "abc@outlo**.com"
#define USER_BAUD_RATE             "460800"
```

4. 修改 CMakeLists.txt

（1）设置本项目的工程名为 pro_4_gimbal_mgmt。

```
# --------设置工程名，将my_project_name 修改为实际工程名--------------
project(pro_4_gimbal_mgmt C)
# ----------------------------------------------------------------
```

（2）添加目录、文件。

```
# -------在此处添加云台管理的目录及该目录下的所有.c 文件----------------------
file(GLOB_RECURSE MODULE_SAMPLE_SRC gimbal_manager/*.c)
include_directories(./gimbal_manager)
# ----------------------------------------------------------------
```

5. 修改 main.c

（1）添加头文件。

```
/*----------在此处按照下面的格式加入云台管理功能的头文件--------------*/
#include <gimbal_manager/gimbal_manager.h>
/*----------------------------------------------------------------*/
```

（2）在 main 函数中设置云台位置并调用云台管理服务入口函数。

```
int main(int argc, char **argv)
{
```

```
    /* 在此处设置云台位置: 1号云台--------------------------*/
    static E_DjiMountPosition mountPosition = DJI_MOUNT_POSITION_PAYLOAD_PORT_NO1;
    /* ---------------------------------------------------*/

    /* 在此处调用云台管理服务入口函数, 使用自由模式-------------*/
    returnCode = DjiTest_GimbalManagerRunSample(
mountPosition,DJI_GIMBAL_MODE_FREE);
    if (returnCode != DJI_ERROR_SYSTEM_MODULE_CODE_SUCCESS) {
        USER_LOG_ERROR("Gimbal Manager init error\n");
    }
    /* ---------------------------------------------------*/
}
```

6. 修改 gimbal_manager.h

头文件无须修改。

7. 修改 gimbal_manager.c

根据 gimbal_manager.c 模板文件，需要完成以下功能。

```
T_DjiReturnCode DjiTest_GimbalManagerRunSample(
                E_DjiMountPosition mountPosition,
                E_DjiGimbalMode gimbalMode){

    /* step 1 初始化云台管理功能-------------------------*/
    returnCode = DjiGimbalManager_Init();
    if (returnCode != DJI_ERROR_SYSTEM_MODULE_CODE_SUCCESS) {
        USER_LOG_ERROR("Init gimbal manager failed, error code: 0x%08X",
returnCode);
        goto out;
    }
    /*------------------------------------------------*/

    /* step 2 设置云台模式------------------------------*/
    returnCode = DjiGimbalManager_SetMode(mountPosition, gimbalMode);
    if (returnCode != DJI_ERROR_SYSTEM_MODULE_CODE_SUCCESS) {
        USER_LOG_ERROR("Set gimbal mode failed, error code: 0x%08X", returnCode);
        goto out;
    }
    /*------------------------------------------------*/

    /* step 3 设置云台回中------------------------------*/
    returnCode = DjiGimbalManager_Reset(mountPosition);
    if (returnCode != DJI_ERROR_SYSTEM_MODULE_CODE_SUCCESS) {
        USER_LOG_ERROR("Reset gimbal failed, error code: 0x%08X", returnCode);
    }
    /*------------------------------------------------*/
```

```c
        USER_LOG_INFO("--> Step 4: Rotate gimbal to target angle by action list\r\n");
        for (int i = 0; i < sizeof(s_rotationActionList) / sizeof(T_DjiTestGimbalActionList); ++i) {
            if (s_rotationActionList[i].action == DJI_TEST_GIMBAL_RESET) {
                USER_LOG_INFO("Target gimbal reset.\r\n");
                returnCode = DjiGimbalManager_Reset(mountPosition);
                if (returnCode != DJI_ERROR_SYSTEM_MODULE_CODE_SUCCESS) {
                    USER_LOG_ERROR("Reset gimbal failed, error code: 0x%08X", returnCode);
                }
                osalHandler->TaskSleepMs(2000);
            } else if (s_rotationActionList[i].action == DJI_TEST_GIMBAL_ROTATION) {

                if (gimbalMode == DJI_GIMBAL_MODE_FREE &&
                    s_rotationActionList[i].rotation.rotationMode == DJI_GIMBAL_ROTATION_MODE_ABSOLUTE_ANGLE) {
                    continue;
                }

                USER_LOG_INFO("Target gimbal pry = (%.1f, %.1f, %.1f)",
                              s_rotationActionList[i].rotation.pitch,
                              s_rotationActionList[i].rotation.roll,
                              s_rotationActionList[i].rotation.yaw);

                rotation = s_rotationActionList[i].rotation;

                /* step 4 设置云台旋转--------------------------------*/
                returnCode = DjiGimbalManager_Rotate(mountPosition, rotation);
                if (returnCode != DJI_ERROR_SYSTEM_MODULE_CODE_SUCCESS) {
                    USER_LOG_ERROR("Target gimbal pry = (%.1f, %.1f, %.1f) failed, error code: 0x%08X",s_rotationActionList[i].rotation.pitch,
s_rotationActionList[i].rotation.roll,s_rotationActionList[i].rotation.yaw,returnCode);
                }
                /* --------------------------------------------------*/
                osalHandler->TaskSleepMs(1000);
            }
        }
    }
```

8. 编译并执行项目

```
# 在项目 pro_4_gimbal_mgmt 目录下，创建 build 目录，并进入 build 目录
pi@raspberrypi:~/ pro_4_gimbal_mgmt $ mkdir build && cd build

# 使用 cmake 创建项目编译环境
pi@raspberrypi:~/ pro_4_gimbal_mgmt /build $ cmake ..
```

```
# make 编译
pi@raspberrypi:~/ pro_4_gimbal_mgmt /build $ make

# 执行文件位于 build/bin 目录下
pi@raspberrypi:~/ pro_4_gimbal_mgmt /build $ cd bin/

# 使用 sudo 执行项目的执行文件
pi@raspberrypi:~/ pro_4_gimbal_mgmt /build/bin $ sudo ./ pro_4_gimbal_mgmt
```

任务评价

任务过程评价表

任务实施人姓名_____ 学号_____ 时间_____

	评价项目及标准	分值	小组评议	教师评议
技术能力	1. 对基本概念的熟悉程度	10		
	2. 掌握云台管理功能接口	10		
	3. 掌握云台管理功能的代码框架	10		
	4. 掌握云台管理模块初始化功能	10		
	5. 掌握使用云台管理功能设置回中的方法	10		
	6. 掌握使用云台管理功能设置旋转的方法	10		
执行能力	1. 出勤情况	5		
	2. 遵守纪律情况	5		
	3. 是否主动参与，有无提问记录	5		
	4. 有无职业意识	5		
社会能力	1. 能否有效沟通	5		
	2. 是否使用基本的文明礼貌用语	5		
	3. 能否与组员主动交流、积极合作	5		
	4. 能否自我学习、自我管理	5		
总分		100		
评定等级：				
评价意见		学习意见		

评定等级具体如下。

A：优，得分>90。B：好，90≥得分>80。C：一般，80≥得分>60。D：有待提高，得分≤60。

小结

本任务首先简要介绍了 DJI 无人机云台管理功能的相关知识；其次介绍了与云台管理功能相关的接口；最后介绍了云台管理功能的代码框架。

项目 5 飞行控制与运动规划

任务 1 飞行控制

🔷 任务要求

PSDK 的无人机飞行控制功能能够设置并获取无人机的各项基础参数,以控制无人机执行基础飞行动作,还能够通过 Joystick 功能控制无人机执行复杂飞行动作。本任务将通过飞行控制功能控制无人机的起飞和降落。

🔷 知识导入

5.1 飞行控制功能

5.1.1 飞行控制概述

PSDK 3.x 版本中的 Joystick 功能为无人机提供了独立于遥控器和 App 的飞行控制能力,该功能在 M300 上为 OSDK 端口功能。机载计算机通过串口向飞行器持续发送控制指令,在水平方向上可支持位置控制模式、速度控制模式、姿态角控制模式、角速度控制模式,而在垂直方向上可支持位置控制模式(相对于起飞点高度)、速度控制模式、油门控制模式。

其基本原理是,机载计算机获取控制权后,通过飞行控制 API 设定飞行模式,并在对应模式下传参,以一定的频率向飞行器发送控制指令。飞行器接收到持续的飞行控制指令后,便按照设定参数进行移动。一旦停止发送指令,飞行器将立即停止移动。此设计可以防止 OSDK 在控制过程中通信突然中断,避免因飞行器路线非预期变化而导致的安全事故。

此功能最大程度地开放了 DJI 无人机的自定义飞行，但是开发工作量会明显增加。区别于航线任务功能，Joystick 功能可以根据开发者的需求脱离飞行控制的定位或导航功能，甚至在无 GPS 的环境下，也可以通过姿态角和油门控制模式，结合开发者的定位导航算法实现飞行控制。但在速度控制模式下，依旧会依赖无人机原有传感器的数据，而在位置控制模式下可以不依赖飞行控制的定位功能。

5.1.2 基础概念

1. 无人机的基础信息

PSDK 开放了设置无人机基础参数的接口，开发者通过设置并获取无人机的基础参数，能够实现对无人机的精准控制。

- 开启或关闭 RTK 功能和避障功能（水平避障和顶部避障）。
- 设置无人机的返航点和返航高度。

具有信息管理功能的应用程序能够方便用户获取无人机基础参数的信息。

- 获取 RTK 功能和避障功能（水平避障和顶部避障）的状态。
- 获取无人机的返航点和返航高度。

2. 无人机的基础飞行动作

无人机的基础飞行动作主要包含无人机锁定、无人机起飞和降落、无人机返航 3 类，开发者使用 PSDK 提供的接口即可根据实际的控制需求，控制无人机执行指定的飞行动作。基础飞行动作如表 5-1 所示。

表 5-1 基础飞行动作

功能类型	基本功能	功能说明
无人机锁定	解锁	无人机解锁后，其螺旋桨会怠速旋转，但不会飞离地面
	上锁	无人机上锁后，其螺旋桨由怠速旋转状态变为静止状态
无人机起飞和降落	自动起飞	在使用该功能时，无人机会自动起飞 1.2 米（该高度不可调整）
	自动降落	在使用该功能时，无人机会自动降落
	取消自动降落	在使用该功能时，无人机在下降过程中会停止降落，并悬停在空中
	降落确认	当无人机降落到距离地面一定高度时，使用该功能可确认无人机是否降落到地面
	强制降落	无视降停面的状态，强制无人机降落（降落速度快）
无人机返航	返航	在使用该功能时，无人机会自动返航
	取消返航	在使用该功能时，无人机会悬停在空中

3. Joystick 功能

Joystick 是一个无人机综合控制功能，在使用 Joystick 功能时，开发者只有根据实际的应用需要，通过调用 Joystick 功能中的接口，同时设置无人机使用的坐标系、水平控制模式、无人机悬停模式、垂直控制的模式、yaw 角度控制模式，才能设计出满足使用需求的无人机飞行控制逻辑。

1）设置坐标系

- 机体坐标系：以无人机的重心为原点，无人机机头前进的方向为 X 轴，机头前进方向的右侧为 Y 轴，Z 轴与 X 轴、Y 轴相互垂直交于重心且指向无人机下方（遵循"右手法则"）。在机体坐标系下，无人机围绕 X 轴、Y 轴和 Z 轴旋转时的飞行动作可称为横滚（无人机仅绕 X 轴旋转）、俯仰（无人机仅绕 Y 轴旋转）和偏航（无人机仅绕 Z 轴旋转）。
- 大地坐标系：也称世界坐标系或当地水平坐标系，在该坐标系下，无人机指向地球正北的方向为 X 轴正方向，指向正东的方向为 Y 轴正方向，X 轴与 Y 轴相互垂直，Z 轴竖直指向无人机下方，在满足"右手法则"的前提下，Z 轴将根据无人机飞行的实际情况调节角度，因此该坐标系还被称为"北东地（NED）坐标系"。

2）设置水平控制模式

- 姿态角控制模式：在该模式下，水平方向的指令用于控制无人机的姿态角（在机体坐标系下，该角度为 roll 和 pitch）。
- 速度控制模式：在该模式下，水平方向的指令用于控制无人机的速度。
- 位置控制模式：在该模式下，水平方向的指令用于控制无人机的位置。
- 角速度控制模式：在该模式下，水平方向的指令用于控制无人机的旋转角速度。

3）设置无人机悬停模式

仅在水平控制模式的速度控制模式下，开发者可以设置无人机悬停模式。

- 开启稳定模式：开启稳定模式后，无人机将在指定的位置上悬停。
- 关闭稳定模式：关闭稳定模式后，无人机将按照速度命令飞行，当无人机的前进速度为 0 时，无人机可能会随风飘动。

4）设置垂直控制模式

- 速度控制模式：控制无人机垂直方向的速度。
- 位置控制模式：控制无人机垂直方向的位置，该位置为相对于起飞点的绝对位置。
- 油门控制模式：控制无人机的油门。

5）设置 yaw 角度控制模式

- 角度控制模式：在该模式下，yaw 方向旋转的指令用于控制 yaw 的角度。
- 角速率控制模式：在该模式下，yaw 方向旋转的指令用于控制 yaw 的角速率。

5.2 接口说明

飞行控制功能的头文件为 dji_flight_control.h，该文件描述了飞行控制功能中结构体和函数原型的关键信息与使用方法。头文件中定义了接口函数原型，如表 5-2 所示。

表 5-2 接口函数原型

功能函数	功能
DjiFlightController_Init	飞行控制模块初始化
DjiFlightController_DeInit	飞行控制模块反初始化
DjiFlightController_SetRtkPositionEnableStatus	开启/关闭 RTK 位置功能
DjiFlightController_GetRtkPositionEnableStatus	获取 RTK 使能状态
DjiFlightController_SetRCLostAction	设置遥控器断连行为
DjiFlightController_GetRCLostAction	获取遥控器断连行为
DjiFlightController_SetHorizontalVisualObstacleAvoidanceEnableStatus	启用/禁用水平视觉避障（向前、向后、向左、向右）
DjiFlightController_GetHorizontalVisualObstacleAvoidanceEnableStatus	获取水平视觉避障（向前、向后、向左、向右）的开关状态
DjiFlightController_SetDownwardsVisualObstacleAvoidanceEnableStatus	启用/禁用向下视觉避障
DjiFlightController_GetDownwardsVisualObstacleAvoidanceEnableStatus	获取向下视觉避障的开关状态
DjiFlightController_ArrestFlying	arrest 飞行
DjiFlightController_CancelArrestFlying	arrest 飞行的退出状态
DjiFlightController_TurnOnMotors	当无人机在地面时开启电机
DjiFlightController_TurnOffMotors	当无人机在地面时关闭电机
DjiFlightController_EmergencyStopMotor	在任何情况下紧急停止电机
DjiFlightController_StartTakeoff	无人机在地面时请求起飞
DjiFlightController_StartLanding	无人机在空中时请求降落
DjiFlightController_CancelLanding	无人机在降落时请求停止降落
DjiFlightController_StartConfirmLanding	在无人机距离地面 0.7m 时确认降落
DjiFlightController_StartForceLanding	在任何情况下强迫降落
DjiFlightController_SetHomeLocationUsingGPSCoordinates	设置自定义的 GPS 返航位置
DjiFlightController_SetHomeLocationUsingCurrentAircraftLocation	使用当前无人机的 GPS（不是 RTK）位置设置返航位置
DjiFlightController_SetGoHomeAltitude	设置返航高度
DjiFlightController_GetGoHomeAltitude	获取返航高度
DjiFlightController_StartGoHome	无人机在空中时请求返航
DjiFlightController_CancelGoHome	无人机在返航时请求停止返航

限于篇幅，以上接口函数原型的具体说明请参阅大疆官网。

5.3 代码模板

1. 代码结构

下面是实现飞行控制功能的代码框架，用户只需要编写用于实现飞行控制功能的 flight_control.c 和 flight_control.h 即可。

```
.
├── 3rdparty
├── CMakeLists.txt
├── dji_init.c
├── dji_init.h
├── flight_control
│   ├── flight_control.c
│   └── flight_control.h
├── hal
├── main.c
├── monitor
├── osal
└── psdk_lib
```

2. 头文件模板

本任务的头文件名为 flight_control.h。该文件定义的内容可满足正常的应用开发需求，无须修改。如果需要实现特殊的功能，则可以按需进行修改。

3. 源文件模板

本任务的源文件名为 flight_control.c。

（1）在飞行控制服务入口函数中完成以下功能。

在 DjiTest_FlightControlRunSample 飞行控制服务入口函数中添加所有需要的功能代码。

```
/* -----在飞行控制服务入口函数中完成以下功能--------------------*/
/* step 1 初始化飞行控制功能--------------------*/
/* step 2 根据传入的控制功能执行控制函数----------*/
/*------------------------------------------*/
T_DjiReturnCode DjiTest_FlightControlRunSample
(E_DjiTestFlightCtrlSampleSelect flightCtrlSampleSelect)
{
    T_DjiReturnCode returnCode;

    USER_LOG_DEBUG("Init flight Control Sample");
    DjiTest_WidgetLogAppend("Init flight Control Sample");

    /* step 1 初始化飞行控制功能--------------------*/

    /*------------------------------------------*/

    /* step 2 根据传入的控制功能执行控制函数----------*/

    /*------------------------------------------*/
```

```c
        USER_LOG_DEBUG("Deinit Flight Control Sample");
        DjiTest_WidgetLogAppend("Deinit Flight Control Sample");
        returnCode = DjiTest_FlightControlDeInit();
        if (returnCode != DJI_ERROR_SYSTEM_MODULE_CODE_SUCCESS) {
            USER_LOG_ERROR("Deinit Flight Control sample failed,error code:0x%08llX", returnCode);
            return returnCode;
        }

        return returnCode;
    }
```

（2）在起飞降落函数中完成以下功能。

```c
/*-----在起飞降落函数中完成以下功能-------------------*/
/* step 1 获取无人机摇杆控制所属控制权 -----------*/
/* step 2 无人机起飞 -----------*/
/* step 3 无人机降落 -------------------------*/
/* step 4 释放无人机摇杆控制所属控制权 -----------*/
/*------------------------------------------------*/

void DjiTest_FlightControlTakeOffLandingSample()
{
    T_DjiReturnCode returnCode;

    USER_LOG_INFO("Flight control takeoff-landing sample start");
    /* step 1 获取无人机摇杆控制所属控制权 -----------*/

    /*------------------------------------------------*/
    s_osalHandler->TaskSleepMs(1000);

/* step 2 无人机起飞 -----------------------------*/

    /*------------------------------------------------*/

    USER_LOG_INFO("Successful take off\r\n");
    s_osalHandler->TaskSleepMs(4000);

/* step 3 无人机降落 -----------------------------*/

    /*------------------------------------------------*/

    USER_LOG_INFO("Successful landing\r\n");
    USER_LOG_INFO("--> Step 4: Release joystick authority");
```

```
    /* step 4 释放无人机摇杆控制所属控制权 ------------*/



    /*----------------------------------------------*/
}
```

❯ 任务设计

本任务基于飞行控制的接口控制无人机的起飞和降落。任务的具体设计要求如下。

（1）从 pro_5_flightcontrol_source_template 项目模板文件中复制一个新项目。项目的工程名为 pro_5_flightcontrol。

（2）本任务将实现飞行控制功能，项目目录下的 flight_control 目录用于编写代码，以实现飞行控制功能 API 接口。

（3）在 flight_control 目录下编写代码，实现飞行控制功能。

① 初始化飞行控制功能。
② 设置飞行控制行为：起飞和降落。
③ 控制无人机摇杆控制所属控制权。
④ 起飞。
⑤ 降落。
⑥ 释放无人机摇杆控制所属控制权。

（4）使用 cmake 编译项目。

（5）执行项目。

❯ 任务实施

限于篇幅，这里只给出任务实施步骤和关键代码。

1. 模板文件

连接树莓派开发板，默认路径为当前登录用户的家目录。查看当前的家目录，该目录下有一个名为 pro_5_flightcontrol_source_template 的目录，此目录下保存了本任务的所有模板文件。

目录结构如下。

```
pi@raspberrypi:~ $ cd pro_5_flightcontrol_source_template/
pi@raspberrypi:~/ pro_5_flightcontrol_source_template $ tree
.
├── 3rdparty
├── CMakeLists.txt
├── dji_init.c
├── dji_init.h
├── flight_control
│   ├── flight_control.c
```

```
|       └── flight_control.h
├── hal
├── main.c
├── monitor
├── osal
└── psdk_lib
```

2. 复制模板目录

将 pro_5_flightcontrol_source_template 目录下的所有文件和目录复制到 pro_5_flightcontrol 目录中。

```
pi@raspberrypi:~ $ cp -r pro_5_flightcontrol_source_template pro_5_flightcontrol
pi@raspberrypi:~ $ cd pro_5_flightcontrol
```

3. 修改 dji_init.c

在 dji_init.c 中填写用户账号信息,开发者需根据实际的 PSDK 用户参数填写。

```
#define USER_App_NAME              "RaspberryPi"
#define USER_App_ID                "126413"
#define USER_App_KEY               "0ea2eadb314d1a6e870975364e2bf4a"
#define USER_App_LICENSE           "HZ/GptycWQ........................"
#define USER_DEVELOPER_ACCOUNT     "abc@outlo**.com"
#define USER_BAUD_RATE             "460800"
```

4. 修改 CMakeLists.txt

(1) 设置本项目的工程名为 pro_5_flightcontrol。

```
# --------设置工程名,将my_project_name修改为实际工程名--------------
project(pro_5_flightcontrol C)
# -----------------------------------------------------------
```

(2) 添加目录、文件。

```
# -------在此处添加飞行控制的目录及该目录下的所有.c文件----------------------
file(GLOB_RECURSE MODULE_SAMPLE_SRC flight_control/*.c)
include_directories(./flight_control)
# -----------------------------------------------------------
```

5. 修改 main.c

(1) 添加头文件。

```
/*----------在此处添加飞行控制的头文件----------------------------*/
#include "flight_control/flight_control.h"
/* -----------------------------------------------------------*/
```

(2) 在 main 函数中调用飞行控制服务入口函数,设置起飞和降落的参数。

```
int main(int argc, char **argv)
{

/*-----在此处添加飞行控制服务入口函数,控制无人机的起飞和落地----------*/
returnCode = DjiTest_FlightControlRunSample(
E_DJI_TEST_FLIGHT_CTRL_SAMPLE_SELECT_TAKE_OFF_LANDING);
```

```
        if (returnCode != DJI_ERROR_SYSTEM_MODULE_CODE_SUCCESS) {
            USER_LOG_ERROR("Flight Control init error\n");
        }
    /* ----------------------------------------------------------*/

    }
```

6. 修改 flight_control.h

头文件无须修改。

7. 修改 flight_control.c

根据 flight_control.c 模板文件,需要完成以下功能。

```
/* -----在飞行控制服务入口函数中完成以下功能---------------------*/
T_DjiReturnCode DjiTest_FlightControlRunSample(
E_DjiTestFlightCtrlSampleSelect flightCtrlSampleSelect)
{
    T_DjiReturnCode returnCode;
    USER_LOG_DEBUG("Init flight Control Sample");

    /* step 1 初始化飞行控制功能--------------------*/
    returnCode = DjiTest_FlightControlInit();
    if (returnCode != DJI_ERROR_SYSTEM_MODULE_CODE_SUCCESS) {
        USER_LOG_ERROR("Init flight Control sample failed,error code:0x%08llX", returnCode);
        return returnCode;
    }
    /* ------------------------------------------- */

    /* step 2 根据传入的控制功能执行控制函数-------------*/
    DjiTest_FlightControlSample(flightCtrlSampleSelect);
    /* ------------------------------------------- */

/*-----在起飞降落函数中完成以下功能---------------------*/

void DjiTest_FlightControlTakeOffLandingSample()
{

    /* step 1 获取无人机摇杆控制所属控制权 -------------*/
    returnCode = DjiFlightController_ObtainJoystickCtrlAuthority();
    if (returnCode != DJI_ERROR_SYSTEM_MODULE_CODE_SUCCESS) {
        USER_LOG_ERROR("Obtain joystick authority failed, error code: 0x%08X", returnCode);
        goto out;
    }
    /*-------------------------------------------------*/
```

```c
    /* step 2 无人机起飞 ------------------------------*/
    if (!DjiTest_FlightControlMonitoredTakeoff()) {
        USER_LOG_ERROR("Take off failed");
        goto out;
    }
    /*----------------------------------------------*/

    /* step 3 无人机降落 ------------------------------*/
    if (!DjiTest_FlightControlMonitoredLanding()) {
        USER_LOG_ERROR("Landing failed");
        goto out;
    }
    /*----------------------------------------------*/

    /* step 4 释放无人机摇杆控制所属控制权 ----------------*/
    returnCode = DjiFlightController_ReleaseJoystickCtrlAuthority();
    if (returnCode != DJI_ERROR_SYSTEM_MODULE_CODE_SUCCESS) {
        USER_LOG_ERROR("Release joystick authority failed, error code: 0x%08X", returnCode);
        goto out;
    }
    /*----------------------------------------------*/

}
```

8. 编译并执行项目

```
# 在项目 pro_5_flightcontrol 目录下，创建 build 目录，并进入 build 目录
pi@raspberrypi:~/pro_5_flightcontrol$ mkdir build && cd build

# 使用 cmake 创建项目编译环境
pi@raspberrypi:~/pro_5_flightcontrol/build $ cmake ..

# make 编译
pi@raspberrypi:~/pro_5_flightcontrol/build $ make

# 执行文件位于 build/bin 目录下
pi@raspberrypi:~/pro_5_flightcontrol/build $cd bin/

# 使用 sudo 执行项目的执行文件
pi@raspberrypi:~/pro_5_flightcontrol/build/bin $ sudo ./pro_5_flightcontrol
```

本任务的程序测试具有一定的危险性。程序一旦运行，无人机就会起飞，因此在进行程序测试时建议卸掉螺旋桨，或者在模拟器中运行。千万不能让装有螺旋桨的无人机在室内运行程序，否则螺旋桨可能会伤人。如果需要运行装有螺旋桨的无人机，则请一定要在室外进行。

任务评价

任务过程评价表

任务实施人姓名_____ 学号_____ 时间_____

	评价项目及标准	分值	小组评议	教师评议
技术能力	1. 对基本概念的熟悉程度	10		
	2. 掌握飞行控制功能接口	10		
	3. 掌握飞行控制功能的代码框架	10		
	4. 掌握飞行控制模块初始化功能	10		
	5. 了解飞行控制功能的危险性	10		
	6. 了解飞行控制功能	10		
执行能力	1. 出勤情况	5		
	2. 遵守纪律情况	5		
	3. 是否主动参与，有无提问记录	5		
	4. 有无职业意识	5		
社会能力	1. 能否有效沟通	5		
	2. 是否使用基本的文明礼貌用语	5		
	3. 能否与组员主动交流、积极合作	5		
	4. 能否自我学习、自我管理	5		
总分		100		
评定等级：				
评价意见		学习意见		

评定等级具体如下。

A：优，得分>90。B：好，90≥得分>80。C：一般，80≥得分>60。D：有待提高，得分≤60。

小结

本任务首先简要介绍了 DJI 无人机飞行控制功能的相关知识；其次介绍了与飞行控制功能相关的接口；最后介绍了飞行控制功能的代码框架。

任务 2　运动规划

任务要求

本任务通过调用 DJI PSDK 的接口，能够控制无人机以指定的高度飞往指定位置并执行相应动作，根据实际的使用需求，还能够控制无人机重复、多次执行指定任务，实现自动化巡航等功能。

知识导入

5.4 运动规划功能

DJI PSDK 为了满足用户对控制无人机飞行自动化的需求，提供了运动规划功能。开发者使用运动规划功能可根据实际的使用需求设计相应的航点任务和热点任务，制定控制无人机自动化飞行的控制逻辑。

5.4.1 航点任务

航点规划是一个控制无人机按照指定的航线飞行，以实现无人机飞行自动化的控制功能。

1. 航点

在使用航点任务时，开发者需要指定航点数量和对应的航点类型。

1）航点数量

在使用航点任务时，开发者需要使用 PSDK 中的 API 接口来指定无人机所需要飞达的航点。DJI PSDK 支持开发者在一个任务中最多添加 65535 个航点，最少添加两个航点。

2）航点类型

航点类型是指无人机在执行航点任务时，飞向该航点的方式，其中包含曲率飞行、直线飞行和协调转弯。

- 曲率飞行：是指当无人机以曲率连续的方式执行飞行任务时，到达指定的航点后不停留，或者在航点处停留；当无人机以曲率不连续的方式执行飞行任务时，在航点处停留。
- 直线飞行：可以是直线进入和直线退出。
- 协调转弯：无人机在到达航点前提前转弯。

2. 动作

在使用航点任务时，开发者或用户可为无人机在指定的航点处添加对应的动作，如拍照、录像或悬停等。

1）动作信息

动作信息主要由动作 ID、动作触发器和动作执行器组成。其中，开发者可指定全局唯一的动作 ID 来标识用户设置的动作；动作触发器是指触发无人机执行动作的触发器，包含多种触发器类型及其对应的参数；动作执行器是指执行用户指定动作的模块，包含多种执行器类型和执行器编号，以及其对应的参数。

2）动作数量

DJI PSDK 支持开发者在一个任务中最多添加 65535 个动作。

3）动作管理

动作管理支持将相机对焦/变焦的动作配置到地面站中自动执行、支持云台角度增量控制、支持配置多个云台和多个相机等。

4）动作触发

如果需要在无人机执行航点任务的过程中，触发无人机执行指定的动作，则需要添加触发该动作的条件，可以是定时触发、距离触发、动作串行触发、动作并行触发等。

3. 速度控制

PSDK 为开发者提供了速度控制功能，能够使开发者为指定的航点配置不同的速度（或为同一个航点配置多个速度），支持开发者在无人机执行航线任务时，修改或查询无人机的全局巡航速度。

4. 断连控制

断连控制为新增的支持配置遥控器断连后继续执行航线任务的功能。需要说明的是，基于 PSDK 开发的应用程序在控制无人机执行任务时，用户可使用遥控器控制无人机的飞行速度、飞行高度和飞行航向等。无人机每次只能执行一个自动化飞行任务，上传新的任务后，已有的飞行任务将会被覆盖。在航点任务中，无人机的航点与无人机的动作没有必然关系，开发者可根据实际情况添加航点和无人机飞行时的动作。

5.4.2 工作流程

航点任务功能按照如下流程控制无人机执行航点任务。

1. 上传航线任务的整体信息

一个航点任务包含航点任务的 ID、航点任务的航点数、任务重复次数、航点任务结束后的动作、最大飞行速度和巡航速度。

2. 上传航点信息

- 基础参数：航点坐标（设置航点的经度、纬度和相对于起飞点的高度）、航点类型、航向类型和飞行速度。
- 可选参数：缓冲距离、航向角度、转向模式、兴趣点、单点最大飞行速度、单点巡航速度。

说明：仅在开发者设置完航点信息的所有基础参数后，才能设置航点信息的可选参数。

3. 设置动作信息（可选）

设置动作 ID、动作触发器和动作执行器。

4. 上传无人机航点任务的信息

5. 控制无人机执行航点任务

上传无人机的航点信息和对应的动作信息后，开发者即可通过指定的接口控制航点任务，如开始、停止或暂停任务，设置或获取巡航速度等。

5.5 Waypoint v2

从宏观功能上看，航线任务是飞行控制的一种，为 OSDK 功能。通过预先上传所有的航点坐标和航点动作功能，启动执行后，飞行器将会按照上传的航点及动作的顺序执行。

与 flight_Control（Joystick）不同，从功能的实现和执行上看，航点任务并非 SDK 的控制功能，SDK 提供了上传航点坐标和动作的功能，以及一系列参数的接口。后续的控制过程由飞行控制执行，SDK 不需要控制权来执行航线任务。所以航线任务的执行过程不是完全由 SDK 来控制的，仅可通过 API 来开始/停止航线任务、暂停/恢复航线任务、设置巡航速度等。

OSDK 功能中的航线任务到目前为止分为 Waypoint mission（v1）、Waypoint v2、Waypoint v3 三个版本。其中，M300 机型仅支持 Waypoint v2，M30 机型仅支持 Waypoint v3。OSDK 3.9 及 OSDK 4.x 版本中的 Waypoint mission（v1）支持其他可用 OSDK 开发的飞行控制或机型。不同版本的差距较大，本任务主要基于 M300 机型介绍 Waypoint v2。

Waypoint v2 提供了上传航线任务、获取和设置巡航速度、监控航线状态和控制航线执行动作等功能。注意：Waypoint v2 目前只支持 M300 RTK 机型。

5.6 接口说明

Waypoint v2 相关功能的头文件为 dji_Waypoint_v2.h，该文件描述了 Waypoint v2 中结构体和函数原型的关键信息与使用方法。头文件中定义了接口函数原型，如表 5-3 所示。

表 5-3　接口函数原型

功能函数	说明
DjiWaypointV2_Init	Waypoint v2 模块初始化
DjiWaypointV2_Deinit	Waypoint v2 模块去初始化
DjiWaypointV2_UploadMission	初始化任务模块设置
DjiWaypointV2_Start	开始任务
DjiWaypointV2_Stop	停止任务
DjiWaypointV2_Pause	暂停任务
DjiWaypointV2_Resume	恢复任务
DjiWaypointV2_GetGlobalCruiseSpeed	从飞行控制中获取全局巡航速度设置
DjiWaypointV2_SetGlobalCruiseSpeed	将全局巡航速度设置为飞行控制
DjiWaypointV2_RegisterMissionEventCallback	使用回调函数订阅 Waypoint v2 事件
DjiWaypointV2_RegisterMissionStateCallback	使用回调函数订阅 Waypoint v2 任务状态

限于篇幅，以上接口函数原型的具体说明请参阅大疆官网。

5.7 代码模板

1. 代码结构

下面是实现运动规划功能的代码框架，用户只需要编写用于实现运动规划功能的 Waypoint_v2.c 和 Waypoint_v2.h 即可。

```
.
├── 3rdparty
├── CMakeLists.txt
├── dji_init.c
├── dji_init.h
├── hal
├── main.c
├── monitor
├── osal
├── psdk_lib
└── Waypoint_v2
    ├── Waypoint_v2.c
    └── Waypoint_v2.h
```

2. 头文件模板

本任务的头文件名为 Waypoint_v2.h。该文件定义的内容可满足正常的应用开发需求，无须修改。如果需要实现特殊的功能，则可以按需进行修改。

3. 源文件模板

本任务的源文件名为 Waypoint_v2.c。

在 DjiTest_WaypointV2RunSample 运动规划服务入口函数中添加所有需要的功能代码。

```
/* ---在运动规划服务入口函数中完成以下功能--------------------------*/

/* step 1 运动规划功能初始化--------------------------------*/
/* step 2 订阅无人机 DJI_FC_SUBSCRIPTION_TOPIC_POSITION_FUSED 数据，无回调函数--*/
/* step 3 注册运动规划回调函数
    /* step 3.1 注册运动规划事件 DjiTest_WaypointV2EventCallback 回调函数 */
    /* step 3.2 注册运动规划状态 DjiTest_WaypointV2StateCallback 回调函数 */
/* step 4 上传运动规划任务----------------------------*/
/* step 5 启动运动规划任务----------------------------*/
/* step 6 设置运动规划巡航速度为 1.5--------------*/
/* step 7 获取运动规划巡航速度--------------------*/
/* step 8 暂停运动规划-------------------------------*/
/* step 9 恢复运动规划-------------------------------*/
```

```c
    /* ----------------------------------------------------------*/

    T_DjiReturnCode DjiTest_WaypointV2RunSample(void)
    {
        T_DjiReturnCode returnCode;
        uint32_t timeOutMs = 1000;
        uint16_t missionNum = 8;
        T_DjiWaypointV2GlobalCruiseSpeed setGlobalCruiseSpeed = 0;
        T_DjiWaypointV2GlobalCruiseSpeed getGlobalCruiseSpeed = 0;

        USER_LOG_INFO("Waypoint v2 sample start");

        USER_LOG_INFO("--> Step 1: Init Waypoint v2 sample");

        /* step 1 运动规划功能初始化----------------------------*/

        /* ---------------------------------------------------*/
        if (returnCode != DJI_ERROR_SYSTEM_MODULE_CODE_SUCCESS) {
            USER_LOG_ERROR("Init Waypoint v2 sample failed, error code: 0x%08X",
returnCode);
            USER_LOG_INFO("Waypoint v2 sample end");
            return returnCode;
        }

        USER_LOG_INFO("--> Step 2: Subscribe gps fused data");

        /* step 2 订阅无人机 DJI_FC_SUBSCRIPTION_TOPIC_POSITION_FUSED 数据，无回调函数--*/

        /* ---------------------------------------------------*/
        if (returnCode != DJI_ERROR_SYSTEM_MODULE_CODE_SUCCESS) {
            USER_LOG_ERROR("Subscribe gps fused data failed, error code: 0x%08X",
returnCode);
            goto out;
        }

        USER_LOG_INFO("--> Step 3: Register Waypoint v2 event and state callback\r\n");

        /* step 3.1 注册运动规划事件 DjiTest_WaypointV2EventCallback 回调函数*/

        /* ----------------------------------------------------------*/
        if (returnCode != DJI_ERROR_SYSTEM_MODULE_CODE_SUCCESS) {
```

```c
        USER_LOG_ERROR("Register Waypoint v2 event failed, error code: 0x%08X", returnCode);
        goto out;
    }
    /* step 3.2 注册运动规划状态 DjiTest_WaypointV2StateCallback 回调函数*/

    /* ----------------------------------------------------------*/
    if (returnCode != DJI_ERROR_SYSTEM_MODULE_CODE_SUCCESS) {
        USER_LOG_ERROR("Register Waypoint v2 state failed, error code: 0x%08X", returnCode);
        goto out;
    }
    osalHandler->TaskSleepMs(timeOutMs);

    USER_LOG_INFO("--> Step 4: Upload Waypoint v2 mission\r\n");
    /* step 4 上传运动规划任务------------------------------------*/

    /* ----------------------------------------------------------*/
    if (returnCode != DJI_ERROR_SYSTEM_MODULE_CODE_SUCCESS) {
        USER_LOG_ERROR("Upload Waypoint v2 mission failed, error code: 0x%08X", returnCode);
        goto out;
    }
    osalHandler->TaskSleepMs(timeOutMs);

    USER_LOG_INFO("--> Step 5: Start Waypoint v2 mission\r\n");
    /* step 5 启动运动规划任务------------------------------------*/

    /* ----------------------------------------------------------*/
    if (returnCode != DJI_ERROR_SYSTEM_MODULE_CODE_SUCCESS) {
        USER_LOG_ERROR("Start Waypoint v2 mission failed, error code: 0x%08X", returnCode);
        goto out;
    }
    osalHandler->TaskSleepMs(20 * timeOutMs);

    USER_LOG_INFO("--> Step 6: Set global cruise speed\r\n");
    /* step 6 设置运动规划巡航速度为 1.5--------------------------*/
    setGlobalCruiseSpeed = 1.5;


    /* ----------------------------------------------------------*/
    if (returnCode != DJI_ERROR_SYSTEM_MODULE_CODE_SUCCESS) {
```

```c
        USER_LOG_ERROR("Set global cruise speed failed, error code: 0x%08X", returnCode);
        goto out;
    }
    osalHandler->TaskSleepMs(timeOutMs);

    USER_LOG_INFO("--> Step 7: Get global cruise speed\r\n");
    /* step 7 获取运动规划巡航速度--------------------------------*/

    /* ----------------------------------------------------------*/
    if (returnCode != DJI_ERROR_SYSTEM_MODULE_CODE_SUCCESS) {
        USER_LOG_ERROR("Get global cruise speed failed, error code: 0x%08X", returnCode);
        goto out;
    }
    USER_LOG_INFO("Current global cruise speed is %f m/s", getGlobalCruiseSpeed);
    osalHandler->TaskSleepMs(timeOutMs);

    USER_LOG_INFO("--> Step 8: Pause Waypoint v2 for 5 s\r\n");
    /* step 8 暂停运动规划----------------------------------------*/

    /* ----------------------------------------------------------*/
    if (returnCode != DJI_ERROR_SYSTEM_MODULE_CODE_SUCCESS) {
        USER_LOG_ERROR("Pause Waypoint v2 failed, error code: 0x%08X", returnCode);
        goto out;
    }
    osalHandler->TaskSleepMs(5 * timeOutMs);

    USER_LOG_INFO("--> Step 9: Resume Waypoint v2\r\n");
    /* step 9 恢复运动规划----------------------------------------*/

    /* ----------------------------------------------------------*/
    if (returnCode != DJI_ERROR_SYSTEM_MODULE_CODE_SUCCESS) {
        USER_LOG_ERROR("Resume Waypoint v2 failed, error code: 0x%08X", returnCode);
        goto out;
    }
    osalHandler->TaskSleepMs(50 * timeOutMs);

    USER_LOG_INFO("--> Step 10: Deinit Waypoint v2 sample\r\n");
out:
    returnCode = DjiTest_WaypointV2DeInit();
    if (returnCode != DJI_ERROR_SYSTEM_MODULE_CODE_SUCCESS) {
```

```
        USER_LOG_ERROR("Deinit Waypoint v2 sample failed, error code: 0x%08X",
returnCode);
    }

    USER_LOG_INFO("Waypoint v2 sample end");

    return returnCode;
}
```

任务设计

本任务基于运动规划的接口控制无人机航点任务。任务的具体设计要求如下。

（1）从 pro_5_Waypoint_v2_source_template 项目模板文件中复制一个新项目。项目的工程名为 pro_5_Waypoint_v2。

（2）本任务将实现运动规划功能，项目目录下的 Waypoint_v2 目录用于编写代码，以实现运动规划功能 API 接口。

（3）在 Waypoint_v2 目录下编写代码，实现运动规划功能。
① 运动规划功能初始化。
② 订阅无人机 DJI_FC_SUBSCRIPTION_TOPIC_POSITION_FUSED 数据。
③ 注册运动规划回调函数。
④ 上传运动规划任务。
⑤ 启动运动规划任务。
⑥ 设置运动规划巡航速度。
⑦ 获取运动规划巡航速度。
⑧ 暂停运动规划。
⑨ 恢复运动规划。

（4）使用 cmake 编译项目。

（5）执行项目。

任务实施

限于篇幅，这里只给出任务实施步骤和关键代码。

1. 模板文件

连接树莓派开发板，默认路径为当前登录用户的家目录。查看当前的家目录，该目录下有一个名为 pro_5_Waypoint_v2_source_template 的目录，此目录下保存了本任务的所有模板文件。

目录结构如下。

```
pi@raspberrypi:~ $ cd pro_5_Waypoint_v2_source_template/
pi@raspberrypi:~/ pro_5_Waypoint_v2_source_template $ tree
.
```

```
├── 3rdparty
├── CMakeLists.txt
├── dji_init.c
├── dji_init.h
├── hal
├── main.c
├── monitor
├── osal
├── psdk_lib
└── Waypoint_v2
    ├── Waypoint_v2.c
└── Waypoint_v2.h
```

2. 复制模板目录

将 pro_5_Waypoint_v2_source_template 目录下的所有文件和目录复制到 pro_5_Waypoint_v2 目录中。

```
pi@raspberrypi:~ $ cp -r pro_5_Waypoint_v2_source_template pro_5_Waypoint_v2
pi@raspberrypi:~ $ cd pro_5_Waypoint_v2/
```

3. 修改 dji_init.c

在 dji_init.c 中填写用户账号信息，开发者需根据实际的 PSDK 用户参数填写。

```
#define USER_App_NAME              "RaspberryPi"
#define USER_App_ID                "126413"
#define USER_App_KEY               "0ea2eadb314d1a6e870975364e2bf4a"
#define USER_App_LICENSE           "HZ/GptycWQ........................"
#define USER_DEVELOPER_ACCOUNT     "abc@outlo**.com"
#define USER_BAUD_RATE             "460800"
```

4. 修改 CMakeLists.txt

（1）设置本项目的工程名为 pro_5_Waypoint_v2。

```
# --------设置工程名，将my_project_name修改为实际工程名----------------
project(pro_5_Waypoint_v2 C)
# ----------------------------------------------------------------
```

（2）添加目录、文件。

```
# --------在此处添加运动规划的目录及该目录下的所有.c文件-----------------
file(GLOB_RECURSE MODULE_SAMPLE_SRC Waypoint_v2/*.c)
include_directories(./Waypoint_v2)
# ----------------------------------------------------------------
```

5. 修改 main.c

（1）添加头文件。

```
/*----------在此处添加运动规划的头文件------------------------*/
#include "Waypoint_v2/Waypoint_v2.h"
/* --------------------------------------------------------*/
```

（2）在 main 函数中调用运动规划服务入口函数。

```c
int main(int argc, char **argv)
{
    /*-----在此处添加运动规划服务入口函数----------------------*/
    returnCode = DjiTest_WaypointV2RunSample();
    if (returnCode != DJI_ERROR_SYSTEM_MODULE_CODE_SUCCESS) {
        USER_LOG_ERROR("Flight Control init error\n");
    }
    /*----------------------------------------------------------*/
}
```

6. 修改 Waypoint_v2.h

头文件无须修改。

7. 修改 Waypoint_v2.c

根据 Waypoint_v2.c 模板文件，需要完成以下功能。

```c
/* ---在运动规划服务入口函数中完成以下功能---------------------------*/
T_DjiReturnCode DjiTest_WaypointV2RunSample(void)
{
    /* step 1 运动规划功能初始化----------------------------*/
    returnCode = DjiTest_WaypointV2Init();
    /* ----------------------------------------------------*/


    /* step 2 订阅无人机DJI_FC_SUBSCRIPTION_TOPIC_POSITION_FUSED 数据，无回调函数-*/
    returnCode = DjiFcSubscription_SubscribeTopic(
    DJI_FC_SUBSCRIPTION_TOPIC_POSITION_FUSED,
            DJI_DATA_SUBSCRIPTION_TOPIC_50_HZ,
    NULL);
    /* ----------------------------------------------------*/


    /* step 3.1 注册运动规划事件 DjiTest_WaypointV2EventCallback 回调函数*/
    returnCode = DjiWaypointV2_RegisterMissionEventCallback(
    DjiTest_WaypointV2EventCallback);
    /* ----------------------------------------------------------*/
    }
    /* step 3.2 注册运动规划状态 DjiTest_WaypointV2StateCallback 回调函数*/
    returnCode = DjiWaypointV2_RegisterMissionStateCallback(
    DjiTest_WaypointV2StateCallback);
    /* ----------------------------------------------------------*/

    /* step 4 上传运动规划任务---------------------------------*/
    returnCode = DjiTest_WaypointV2UploadMission(missionNum);
    /* ----------------------------------------------------*/
```

```
    /* step 5 启动运动规划任务------------------------------------*/
    returnCode = DjiWaypointV2_Start();
    /* ----------------------------------------------------------*/

    /* step 6 设置运动规划巡航速度为 1.5-----------------------------*/
    setGlobalCruiseSpeed = 1.5;
returnCode = DjiWaypointV2_SetGlobalCruiseSpeed(
setGlobalCruiseSpeed);
    /* ----------------------------------------------------------*/

    /* step 7 获取运动规划巡航速度---------------------------------*/
returnCode = DjiWaypointV2_GetGlobalCruiseSpeed(
&getGlobalCruiseSpeed);
    /* ----------------------------------------------------------*/

    /* step 8 暂停运动规划----------------------------------------*/
    returnCode = DjiWaypointV2_Pause();
    /* ----------------------------------------------------------*/

    /* step 9 恢复运动规划----------------------------------------*/
    returnCode = DjiWaypointV2_Resume();
    /* ----------------------------------------------------------*/
}
```

8. 编译并执行项目

```
# 在项目 pro_5_Waypoint_v2 目录下，创建 build 目录，并进入 build 目录
pi@raspberrypi:~/ pro_5_Waypoint_v2$ mkdir build && cd build

# 使用 cmake 创建项目编译环境
pi@raspberrypi:~/ pro_5_Waypoint_v2/build $ cmake ..

# make 编译
pi@raspberrypi:~/ pro_5_Waypoint_v2/build $ make

# 执行文件位于 build/bin 目录下
pi@raspberrypi:~/ pro_5_Waypoint_v2/build $cd bin/

# 使用 sudo 执行项目的执行文件
pi@raspberrypi:~/ pro_5_Waypoint_v2/build/bin $ sudo ./ pro_5_Waypoint_v2
```

本任务的程序测试具有一定的危险性。程序一旦运行，无人机就会起飞，因此在进行程序测试时建议卸掉螺旋桨，或者在模拟器中运行。千万不能让装有螺旋桨的无人机在室内运行程序，否则螺旋桨可能会伤人。如果需要运行装有螺旋桨的无人机，则请一定要在室外进行。

任务评价

任务过程评价表

任务实施人姓名_____ 学号_____ 时间_____

评价项目及标准		分值	小组评议	教师评议
技术能力	1. 对基本概念的熟悉程度	10		
	2. 掌握 Waypoint_v2 接口	10		
	3. 掌握 Waypoint_v2 的代码框架	10		
	4. 掌握 Waypoint_v2 初始化功能	10		
	5. 了解 Waypoint_v2 的危险性	10		
	6. 了解 Waypoint_v2 的其他控制功能	10		
执行能力	1. 出勤情况	5		
	2. 遵守纪律情况	5		
	3. 是否主动参与，有无提问记录	5		
	4. 有无职业意识	5		
社会能力	1. 能否有效沟通	5		
	2. 是否使用基本的文明礼貌用语	5		
	3. 能否与组员主动交流、积极合作	5		
	4. 能否自我学习、自我管理	5		
总分		100		
评定等级：				
评价意见		学习意见		

评定等级具体如下。

A：优，得分>90。B：好，90≥得分>80。C：一般，80≥得分>60。D：有待提高，得分≤60。

小结

本任务首先简要介绍了 DJI 无人机运动规划功能的相关知识；其次介绍了与 Waypoint v2 相关的接口；最后介绍了 Waypoint v2 的代码框架。

项目 6

自定义控件

➲ 任务要求

本项目将通过 DJI PSDK 的 App 窗口功能组件自定义控件，实现在 DJI Pilot UI 上获取负载设备的系统时间。

➲ 知识导入

6.1 App 窗口功能组件

DJI 无人机的 App 窗口功能组件也被称为 Widget 控件，是一个将"负载设备的功能"封装为按钮、开关、滑动条、选择列表、输入框等控件的功能。使用 DJI Pilot 或基于 MSDK 开发的移动端 App 能够识别负载设备中控件的配置信息并生成 UI 控件，方便用户快速设置负载设备的参数并控制负载设备执行指定动作。同时，使用 DJI Pilot 或基于 MSDK 开发的移动端 App 还能够以浮窗的形式显示负载设备的状态信息。此外，用户能够根据使用需求，将负载设备的功能映射到遥控器的预留按键上，通过使用遥控器的预留按键，能够以更便捷的方式控制负载设备。

Widget 功能定义了按钮、开关、滑动条、选择列表、输入框等控件，以及主界面浮窗、设置界面文本框。不同的控件对应不同的状态值，如按钮对应按下和释放，滑动条对应可变范围值。

当使用 App 对上述几种类型的控件进行操作时，对应控件的定义状态会发生变化，PSDK 通过捕获控件状态的变更就可以感知 App 的操作。PSDK 端与 DJI Pilot 具体是怎样传输数据的，开发者不需要关注，实际上，在 DJI Pilot 中开发者也无法访问底层的数据传输。开发者只需要关注这几个控件分别对应什么状态，并将所需功能按照这几类控件的状态进行封装处理即可，通过 PSDK 注册就完成了负载端的 Widget 控件实现。常用控件及其状态变化如表 6-1 所示。

表 6-1 常用控件及其状态变化

控件类型	状态变化
BUTTON：按钮	按下（Press）/释放（Release）
SWITCH：开关	开（ON）/关（OFF）
SCALE：滑动条	0%～100%的可变值，滑动条状态表示的是一个范围的百分比（0%～100%）。例如，负载端需要一个关于音量的滑动条
LIST：选择列表	包含一系列预设可选项的选择列表，使用整数值对应可选项，取值范围为 0～$N-1$，N 表示设置的列表项的个数
INT_INPUT_BOX：整数输入框	该控件是一个整数输入框，其中只能输入整数。区别于文本输入框，文本输入框不需要在 PSDK 中定义和注册，与主界面浮窗文本一样，直接对应 API

6.2 自定义控件 UI

Widget 控件是 UI 概念，在 PSDK 中对应的是 DJI Pilot 端的 UI 控件。PSDK/OSDK 作为机载端开发的负载设备，在通常情况下，用户无法直接与负载设备进行交互和控制，当然，开发者可以通过 SSH 等远程手段进入机载设备进行查看或调试，但更多的终端用户是通过遥控器 App 进行可视化操控的。Widget 控件可以使 PSDK/OSDK 中的一些用户交互感知功能在 App 的 UI 上呈现出来，这样用户在使用遥控器 App 和 DJI Pilot 时，就可以对集成的第三方负载进行控制或设置。

Widget 控件根据显示区域的不同，分为主界面控件和配置界面控件。二者最大的区别在于主界面控件图标是立即可见的，并且可以自定义图标展示。

6.2.1 主界面控件

主界面控件分为以下两类。
- 动作栏控件：动作栏控件支持按钮、开关、滑动条和选择列表 4 种控件类型，最多支持设置 5 个自定义控件。
- 浮窗：实时显示负载设备的状态信息。

主界面控件显示在负载的图传画面之上（如果不实现图传，则是一个纯黑的界面）。注意：如果主界面控件不能显示，则可在启用 camera_emu 模块相机功能后继续尝试。

6.2.2 配置界面控件

配置界面控件是指用户能够操作配置界面的控件，如按钮、开关、滑动条、选择列表、文本输入框、整数输入框。

用户可在主界面中点击负载图标，或者通过菜单栏中的负载栏进入配置界面。相较于主界面控件，配置界面控件不能使用自定义 UI，且控件类型增加了文本输入框和整数输入框。其中，最多支持设置 1 个文本输入框，最多允许输入 128 个 UTF-8 字符，文本字符串通过数据传输接口被发送到负载端。受日志长度的影响，用户所输入的字符可能不会全部显示出来。

6.2.3 控件配置文件

控件 UI 的展示、编码开发功能的实现都依赖控件配置文件，控件配置文件为不含注释的 JSON 文件。对于主界面的动作栏控件，控件配置文件提供了控件 UI 图标。DJI Pilot 系统语言支持中、英文。

1. 文件路径

控件配置文件路径下包含静态配置文件（JSON 文件）和控件 UI 图标文件（PNG 文件），控件配置文件的路径为 Widget/Widget_file/，该目录结构如下。

```
pi@raspberrypi:~/pro_6_Widget/Widget $ tree Widget_file
Widget_file
├── cn_big_screen                    # 当系统语言为中文时使用 cn_big_screen
│   ├── icon_button1.png             # *.png 文件是 UI 图标
│   ├── icon_button2.png
│   ├── icon_list_item1.png
│   ├── icon_list_item2.png
│   ├── icon_scale.png
│   ├── icon_switch_select.png
│   ├── icon_switch_unselect.png
│   └── Widget_config.json           # 当系统语言为中文时使用的控件配置文件
└── en_big_screen                    # 当系统语言为英文时使用 en_big_screen
    ├── icon_button1.png             # *.png 文件是 UI 图标
    ├── icon_button2.png
    ├── icon_list_item1.png
    ├── icon_list_item2.png
    ├── icon_scale.png
    ├── icon_switch_select.png
    ├── icon_switch_unselect.png
    └── Widget_config.json           # 当系统语言为英文时使用的控件配置文件
```

当 DJI Pilot 系统语言为中文时，控件配置文件为 cn_big_screen；当 DJI Pilot 系统语言为英文时，控件配置文件为 en_big_screen。不同系统语言下的配置信息如控件编号、数量与类型等需要保持一致。

2. 配置控件属性

Widget_config.json 是一个用于配置控件静态属性的文件，在修改 Widget_config.json 文件时，请务必严格遵守 JSON 的语法规则，否则配置文件将无法使用。建议先在静态配置文件中配置控件属性，再设计控件 UI 图标。

关于 JSON 文件格式，要注意以下几点。
- JSON 文件中的配置项需要包裹在"{}"中，通过 key-value 的方式表达数据。
- JSON 的 key 必须包裹在""""中，请勿丢失 key 外部的""""。
- JSON 的 value 只支持数字（含浮点数和整数）、字符串、Bool 值（如 true 和 false）、数组（需要包裹在"[]"中）和对象（需要包裹在"{}"中）。

下面给出了控件 UI 示例中的静态配置文件。代码中"//"后的内容为代码注释，在实际的静态配置文件中请勿添加该内容。

```
{
    "version": {                        // 自定义控件配置文件版本，用户不可更改此版本号
        "major" : 1,
        "minor" : 0
    },
    "main_interface": {                 // DJI Pilot 主界面控件设置
        "floating_window": {            // 浮窗配置
            "is_enable": true           // 浮窗是否显示，true 表示显示，false 表示隐藏
        },
        "Widget_list": [                // 主界面动作栏控件列表
            {
                "Widget_index": 0,              // 控件编号
                "Widget_type": "button",        // 控件类型，主界面 Action 控件支持
"button"表示按钮, "switch"表示开关, "range"表示滑动条, "list"表示选择列表
                "Widget_name": "Button_1",      // 控件名称
                "icon_file_set": {              // 控件图标文件集
                    "icon_file_name_selected" : "icon_button1.png",  // 选中状态下的控件图标文件名称
                    "icon_file_name_unselected" : "icon_button1.png" // 未选中状态下的控件图标文件名称
                }
            },
            {
                "Widget_index": 1,
                "Widget_type": "button",
                "Widget_name": "Button_2",
                "icon_file_set": {
                    "icon_file_name_selected" : "icon_button2.png",
                    "icon_file_name_unselected" : "icon_button2.png"
                }
            },
            {
                "Widget_index": 2,
                "Widget_type": "list",
                "Widget_name": "List",
                "list_item": [
```

```json
                {
                    "item_name": "Item_1",
                    "icon_file_set": {
                        "icon_file_name_selected" : "icon_list_item1.png",
                        "icon_file_name_unselected" : "icon_list_item1.png"
                    }
                },
                {
                    "item_name": "Item_2",
                    "icon_file_set": {
                        "icon_file_name_selected" : "icon_list_item2.png",
                        "icon_file_name_unselected" : "icon_list_item2.png"
                    }
                }
            ]
        },
        {
            "Widget_index": 3,
            "Widget_type": "switch",
            "Widget_name": "Switch",
            "icon_file_set": {
                "icon_file_name_selected" : "icon_switch_select.png",
                "icon_file_name_unselected" : "icon_switch_unselect.png"
            }
        },
        {
            "Widget_index": 4,
            "Widget_type": "scale",
            "Widget_name": "Scale",
            "icon_file_set": {
                "icon_file_name_selected" : "icon_scale.png",
                "icon_file_name_unselected" : "icon_scale.png"
            }
        }
    ]
},
"config_interface": {
    "text_input_box": {                      // 文本输入框
        "Widget_name":"TextInputBox",        //文本输入框名称
        "placeholder_text":"Please input message",    //文本输入框占位符文本
        "is_enable":false     // 文本输入框是否显示，false表示不显示，true表示显示
    },
    "Widget_list": [
        {
            "Widget_index": 5,
            "Widget_type": "switch",
```

```
            "Widget_name": "Red LED"
        },
        {
            "Widget_index": 6,
            "Widget_type": "switch",
            "Widget_name": "Blue LED"
        },
        {
            "Widget_index": 7,
            "Widget_type": "button",
            "Widget_name": "Button_1"
        },
        {
            "Widget_index": 8,
            "Widget_type": "scale",
            "Widget_name": "Scale_1"
        },
        {
            "Widget_index": 9,
            "Widget_type": "int_input_box",
            "Widget_name": "InputBox_1",
            "int_input_box_hint": "unit:m"
        },
    ]
  }
}
```

6.3 接口说明

自定义控件相关功能的头文件为 dji_Widget.h，该文件描述了自定义控件功能中结构体和函数原型的关键信息与使用方法。主要函数说明如下。

1. 初始化自定义控件模块

在使用自定义控件功能时，请先初始化自定义控件模块。

```
T_DjiReturnCode DjiWidget_Init(void);
```

2. 注册自定义控件配置文件的默认路径

使用 DjiWidget_RegDefaultUiConfigByDirPath 函数注册自定义控件配置文件的默认路径。

```
T_DjiReturnCode DjiWidget_RegDefaultUiConfigByDirPath(
const char *WidgetConfigDirPath);
```

说明：参数 WidgetConfigDirPath 是自定义控件配置文件的路径。本函数仅支持基于 Linux 系统开发的负载设备使用。

3. 注册自定义控件配置文件的路径

使用 DjiWidget_RegUiConfigByDirPath 函数注册自定义控件配置文件的路径。

```
T_DjiReturnCode DjiWidget_RegUiConfigByDirPath(
E_DjiMobileAppLanguage appLanguage,
                E_DjiMobileAppScreenType appScreenType,
                const char *WidgetConfigDirPath);
```

说明：参数 appLanguage 是基于 MSDK 开发的移动端 App 的语言类型；参数 appScreenType 是终端设备的屏幕类型；参数 WidgetConfigDirPath 是自定义控件配置文件的地址。本函数仅支持基于 Linux 系统开发的负载设备使用，不同语言和屏幕尺寸下的控件配置需要有相同的控件类型、索引和数量。

4. 注册二进制配置文件的默认路径

使用 DjiWidget_RegDefaultUiConfigByBinaryArray 函数注册二进制配置文件的默认路径。

```
T_DjiReturnCode DjiWidget_RegDefaultUiConfigByBinaryArray(
const T_DjiWidgetBinaryArrayConfig *binaryArrayConfig
);
```

说明：参数 binaryArrayConfig 是二进制数组控件配置文件。

5. 通过二进制数组控件文件注册控件配置信息

使用 DjiWidget_RegUiConfigByBinaryArray 函数通过二进制数组控件文件注册控件配置信息。

```
T_DjiReturnCode DjiWidget_RegUiConfigByBinaryArray(
E_DjiMobileAppLanguage appLanguage,
                E_DjiMobileAppScreenType screenType,
                const T_DjiWidgetBinaryArrayConfig *binaryArrayConfig);
```

说明：参数 appLanguage 是基于 MSDK 开发的移动端 App 的语言类型；参数 screenType 是终端设备的屏幕类型；参数 binaryArrayConfig 是二进制数组控件配置文件。

6. 注册自定义控件处理函数列表

使用 DjiWidget_RegHandlerList 函数注册自定义控件处理函数列表。

```
T_DjiReturnCode DjiWidget_RegHandlerList(
const T_DjiWidgetHandlerListItem *WidgetHandlerList,
uint32_t itemCount);
```

说明：参数 WidgetHandlerList 是自定义控件处理函数列表；参数 itemCount 是列表项数。

7. 向移动端 App 的浮窗发送消息

使用 DjiWidgetFloatingWindow_ShowMessage 函数向 DJI Pilot 或基于 MSDK 开发的移动端 App 的浮窗发送消息。

```
T_DjiReturnCode DjiWidgetFloatingWindow_ShowMessage(const char *str);
```

说明：参数 str 是指向需要向 DJI Pilot 或基于 MSDK 开发的移动端 App 的浮窗发送的消息，其长度不可超过 PSDK_WIDGET_FLOATING_WINDOW_MSG_MAX_LEN 中设置的值。向 DJI Pilot 或基于 MSDK 开发的移动端 App 的浮窗发送的消息的最大带宽为 2KB/s。

8. 获取消息发送通道的状态

使用 DjiWidgetFloatingWindow_GetChannelState 函数获取消息发送通道的状态。

```
T_DjiReturnCode DjiWidgetFloatingWindow_GetChannelState(
T_DjiDataChannelState *state);
```

说明：参数 state 是指向 DJI Pilot 或基于 MSDK 开发的移动端 App 的浮窗发送消息的通道的状态。

6.4 模板代码

开发者在实现自定义控件功能时，不要在 main 函数中调用自定义控件的接口，而需要创建用户线程来实现自定义控件功能。在用户线程中调用自定义控件的接口，启动调度器后，自定义控件接口将正常运行。

1. 代码结构

下面是实现自定义控件功能的代码框架，用户只需要编写用于实现自定义控件功能的 Widget.c 和 Widget.h 即可。

```
├── 3rdparty            // 第三方库文件目录，省略了该目录下的内容
├── CMakeLists.txt      // cmake 构建项目的文件
├── dji_init.c          // 系统初始化源文件
├── dji_init.h          // 系统初始化头文件
├── Widget              // 用户实现自定义控件的目录
│    ├── Widget.c       // 实现消息自定义控件的源文件
│    └── Widget.h       // 头文件
├── hal                 // hal 文件目录，省略了该目录下的内容
├── main.c              // 主函数文件
├── monitor             // 监控文件目录，省略了该目录下的内容
├── osal                // osal 文件目录，省略了该目录下的内容
└── psdk_lib            // psdk 库文件目录，省略了该目录下的内容
```

2. 头文件模板

本项目中的头文件名为 Widget.h。该文件中定义的内容可满足正常的应用开发需求，如果需要实现特殊的功能，则可以按需要修改。

在头文件中定义了一个 DjiTest_WidgetStartService 函数，该函数在 main.c 中被调用，作为自定义控件的入口函数。本项目中的头文件 Widget.h 无须修改，这里不再列出。

3. 源文件模板

本项目的源文件名为 Widget.c，以下是 Widget.c 的模板文件，按照提示即可完成代码的编写。

```
#include <dji_Widget.h>
#include <dji_logger.h>
```

```c
#include "utils/util_misc.h"
#include <dji_platform.h>
#include <stdio.h>
#include "file_binary_array_list_en.h"

#include "Widget.h"

/* Private constants ---------------------------------------------------------*/
#define WIDGET_DIR_PATH_LEN_MAX            (256)
#define WIDGET_TASK_STACK_SIZE             (2048)

/* Private types -------------------------------------------------------------*/

/* Private functions declaration ---------------------------------------------*/
static void *DjiTest_WidgetTask(void *arg);
static T_DjiReturnCode DjiTestWidget_SetWidgetValue(E_DjiWidgetType WidgetType, uint32_t index, int32_t value,
                                                    void *userData);
static T_DjiReturnCode DjiTestWidget_GetWidgetValue(E_DjiWidgetType WidgetType, uint32_t index, int32_t *value,
                                                    void *userData);

/* Private values ------------------------------------------------------------*/
static T_DjiTaskHandle s_WidgetTestThread;
static bool s_isWidgetFileDirPathConfigured = false;
static char s_WidgetFileDirPath[DJI_FILE_PATH_SIZE_MAX] = {0};

static const uint32_t s_WidgetHandlerListCount = sizeof(s_WidgetHandlerList) / sizeof(T_DjiWidgetHandlerListItem);
static int32_t s_WidgetValueList[sizeof(s_WidgetHandlerList) / sizeof(T_DjiWidgetHandlerListItem)] = {0};

/* 在 Widget 服务入口函数 DjiTest_WidgetStartService 中完成以下任务--------*/
/* step 1  调用 DjiWidget_Init 函数初始化自定义控件功能----------------*/
/* step 2  调用 DjiWidget_RegUiConfigByDirPath 函数指定适配语言和屏幕大小*/
/* step 3  调用 DjiWidget_RegHandlerList 函数注册自定义控件处理函数列表--*/
/* step 4  调用 osalHandler->TaskCreate 函数，启动线程（任务）来使 Widget 工作，线程名
称为 DjiTest_WidgetTask-----*/

/* ---------------------------------------------------------------*/
T_DjiReturnCode DjiTest_WidgetStartService(void)
{
    T_DjiReturnCode djiStat;
    T_DjiOsalHandler *osalHandler = DjiPlatform_GetOsalHandler();

    /* 在 Widget 服务入口函数 DjiTest_WidgetStartService 中完成以下任务----*/
```

```
        /* step 1  调用 DjiWidget_Init 函数初始化自定义控件功能----------*/
        /* step 2  调用 DjiWidget_RegUiConfigByDirPath 函数指定适配语言和屏幕大小-----*/
        /* step 3  调用 DjiWidget_RegHandlerList 函数注册自定义控件处理函数列表--------*/
        /* step 4  调用 osalHandler->TaskCreate 函数，启动线程（任务）来使 Widget 工作，线
程名称: DjiTest_WidgetTask-----*/

    /* ------------------------------------------------------------*/

    return DJI_ERROR_SYSTEM_MODULE_CODE_SUCCESS;
}

__attribute__((weak)) void DjiTest_WidgetLogAppend(const char *fmt, ...)
{

}

static void *DjiTest_WidgetTask(void *arg)
{
    char message[DJI_WIDGET_FLOATING_WINDOW_MSG_MAX_LEN];
    uint32_t sysTimeMs = 0;
    T_DjiReturnCode djiStat;
    T_DjiOsalHandler *osalHandler = DjiPlatform_GetOsalHandler();

    USER_UTIL_UNUSED(arg);

    while (1) {
        /* -----此处完成使 Widget 工作的线程功能----------*/
        /* step 1 调用 GetTimeMs 函数获取当前系统时间--------------------*/
        /* step 2 调用 snprintf 函数拼接输出消息-----------------------*/
        /* step 3 此处调用 DjiWidgetFloatingWindow_ShowMessage 函数向 DJI Pilot 的
浮窗发送消息------*/

        /* ------------------------------------------------------ */

        osalHandler->TaskSleepMs(1000);
    }
}
```

⮕ 任务设计

在通常情况下，用户无法直接与负载设备进行交互和控制，Widget 控件可以将 PSDK 中的一些用户交互感知功能在 App 的 UI 上呈现出来，这样用户在使用遥控器和 DJI Pilot 时，可以对集成的第三方负载进行控制或设置。本项目将通过 DJI PSDK 的 App 窗口功能组件自定义控件，实现在 DJI Pilot UI 上获取负载设备的系统时间。

任务的具体设计要求如下。
(1) 创建项目目录。
(2) 复制 psdk_lib、hal、osal、monitor、3rdparty、main.c 和 CMakeLists.txt 到该项目目录下。
(3) 本项目将实现自定义控件功能，项目目录下的 Widget 目录用于编写代码，以实现用户线程调用自定义控件 API 接口。
(4) 在 Widget 目录下编写代码，实现消息订阅的自定义控件。
① 初始化自定义控件功能。
② 指定适配自定义控件的语言和屏幕大小。
③ 注册自定义控件处理函数列表。
④ 启动线程（任务）来使 Widget 工作。
⑤ 在用户线程中获取负载设备的系统时间。
(5) 使用 cmake 编译项目。
(6) 执行项目。

➡ 任务实施

限于篇幅，这里只给出任务实施步骤和关键代码。

1. 模板文件

连接树莓派开发板，默认路径为当前登录用户的家目录。查看当前的家目录，该目录下有一个名为 pro_6_source_template 的目录，此目录下保存了本项目的所有模板文件。

目录结构如下。

```
pi@raspberrypi:~ $ cd pro_6_source_template/
pi@raspberrypi:~/pro_6_source_template $ tree
.
├── 3rdparty              // 第三方库文件目录，省略了该目录下的内容
├── CMakeLists.txt        // cmake 构建项目的文件
├── dji_init.c            // 系统初始化源文件
├── dji_init.h            // 系统初始化头文件
├── Widget                // 用户实现自定义控件的目录
│   ├── Widget.c          // 实现消息自定义控件的源文件
│   └── Widget.h          // 头文件
├── hal                   // hal 文件目录，省略了该目录下的内容
├── main.c                // 主函数文件
├── monitor               // 监控文件目录，省略了该目录下的内容
├── osal                  // osal 文件目录，省略了该目录下的内容
└── psdk_lib
```

2. 复制模板目录

将 pro_6_source_template 目录下的所有文件和目录都复制到 pro_6_Widget 目录中。

```
pi@raspberrypi:~ $ cp -r pro_6_source_template pro_6_Widget
```

3. 修改 dji_init.c

在 dji_init.c 中填写用户账号信息，开发者需根据实际的 PSDK 用户参数填写。

```
#define USER_App_NAME                    "RaspberryPi"
#define USER_App_ID                      "126413"
#define USER_App_KEY                     "0ea2eadb314d1a6e870975364e2bf4a"
#define USER_App_LICENSE                 "HZ/GptycWQ........................"
#define USER_DEVELOPER_ACCOUNT           "abc@outlo**.com"
#define USER_BAUD_RATE                   "460800"
```

4. 修改 CMakeLists.txt

在 CMakeLists.txt 中需要修改 3 处，分别是修改项目的工程名、添加自定义控件的目录和文件。按照以下提示找到相应的位置进行修改即可。

（1）设置本项目的工程名为 pro_6_Widget。

```
# -----------------------设置工程名------------------------------------
project(pro_6_Widget C)
# --------------------------------------------------------------------
```

（2）添加目录、文件。

```
# ---在此处添加自定义控件的文件 Widget/*.c 和目录 ./Widget-------------------
file(GLOB_RECURSE MODULE_SAMPLE_SRC Widget/*.c)
include_directories(./Widget)
# --------------------------------------------------------------------
```

修改后保存并退出。

5. 修改 main.c

在 main.c 中的 main 函数内部完成下面的功能。

```
/* --------此处调用 DjiTest_WidgetStartService 函数初始化自定义控件服务-*/
returnCode = DjiTest_WidgetStartService();
if (returnCode != DJI_ERROR_SYSTEM_MODULE_CODE_SUCCESS) {
    USER_LOG_ERROR("Widget sample init error");
}
/* ----------------------------------------------------------------*/
```

6. 修改 Widget.h

自定义控件头文件 Widget.h 无须修改。

7. 修改 Widget.c

根据 Widget.c 的模板文件，需要完成两个函数的功能。

在 DjiTest_WidgetStartService 函数中添加相应的代码，完成后的代码如下。

```
/* 在 Widget 服务入口函数 DjiTest_WidgetStartService 中完成以下功能--------*/
T_DjiReturnCode DjiTest_WidgetStartService(void)
{
    /* step 1  调用 DjiWidget_Init 函数初始化自定义控件功能-----*/
    djiStat = DjiWidget_Init();
```

```c
    if (djiStat != DJI_ERROR_SYSTEM_MODULE_CODE_SUCCESS) {
        USER_LOG_ERROR("Dji test Widget init error, stat = 0x%08llX", djiStat);
        return djiStat;
    }
     /* -----------------------------------------------------*/

    /* step 2  调用 DjiWidget_RegUiConfigByDirPath 函数指定适配语言和屏幕大小-----*/
    char *tempPath = "../../Widget/Widget_file/cn_big_screen";
    djiStat = DjiWidget_RegUiConfigByDirPath(DJI_MOBILE_App_LANGUAGE_CHINESE,
                    DJI_MOBILE_App_SCREEN_TYPE_BIG_SCREEN,
                                            tempPath);
    if (djiStat != DJI_ERROR_SYSTEM_MODULE_CODE_SUCCESS) {
        USER_LOG_ERROR("Add Widget ui config error, stat = 0x%08llX", djiStat);
        return djiStat;
    }
     /* -------------------------------------------------------------------*/

    /* step 3  调用 DjiWidget_RegHandlerList 函数注册自定义控件处理函数列表-----*/
    djiStat = DjiWidget_RegHandlerList(s_WidgetHandlerList,
s_WidgetHandlerListCount);
    if (djiStat != DJI_ERROR_SYSTEM_MODULE_CODE_SUCCESS) {
        USER_LOG_ERROR("Set Widget handler list error, stat = 0x%08llX", djiStat);
        return djiStat;
    }
     /* -------------------------------------------------------------------*/

    /* step 4  调用 osalHandler->TaskCreate 函数，启动线程（任务）来使 Widget 工作，线
程名称为 DjiTest_WidgetTask-----*/
    if (osalHandler->TaskCreate("user_Widget_task", DjiTest_WidgetTask,
                                            WIDGET_TASK_STACK_SIZE,
    NULL,&s_WidgetTestThread) != DJI_ERROR_SYSTEM_MODULE_CODE_SUCCESS) {
        USER_LOG_ERROR("Dji Widget test task create error.");
        return DJI_ERROR_SYSTEM_MODULE_CODE_UNKNOWN;
    }
     /* -------------------------------------------------------------------*/
}
/*------在 DjiTest_WidgetTask 函数中添加相应的代码------------------*/

static void *DjiTest_WidgetTask(void *arg)
{
    while (1) {

        /* step 1 调用 GetTimeMs 函数获取当前系统时间--------*/
```

```
            djiStat = osalHandler->GetTimeMs(&sysTimeMs);
            if (djiStat != DJI_ERROR_SYSTEM_MODULE_CODE_SUCCESS) {
                USER_LOG_ERROR("Get system time ms error, stat = 0x%08llX", djiStat);
            }
        /* ------------------------------------------*/


        /* step 2 调用 snprintf 函数拼接输出消息--------------*/
            snprintf(message, DJI_WIDGET_FLOATING_WINDOW_MSG_MAX_LEN, "System time : %u ms", sysTimeMs);
        /* ------------------------------------------*/


        /* step 3 调用 DjiWidgetFloatingWindow_ShowMessage 函数向 DJI Pilot 的浮窗
发送消息--------------------*/
            djiStat = DjiWidgetFloatingWindow_ShowMessage(message);
            if (djiStat != DJI_ERROR_SYSTEM_MODULE_CODE_SUCCESS) {
                USER_LOG_ERROR("Floating window show message error, stat = 0x%08llX", djiStat);
            }
        /* ------------------------------------------*/


            osalHandler->TaskSleepMs(1000);
        }
    }
```

8. 编译并执行项目

```
# 在项目 pro_6_Widget 目录下, 创建 build 目录, 并进入 build 目录
pi@raspberrypi:~/pro_6_Widget$ mkdir build && cd build

# 使用 cmake 创建项目编译环境
pi@raspberrypi:~/pro_6_Widget/build $ cmake ..

# make 编译
pi@raspberrypi:~/pro_6_Widget/build $ make

# 执行文件位于 build/bin 目录下
pi@raspberrypi:~/pro_6_Widget/build $cd bin/

# 使用 sudo 命令执行项目的执行文件
pi@raspberrypi:~/pro_6_Widget/build/bin $ sudo ./pro_6_Widget
```

任务评价

任务过程评价表

任务实施人姓名_____ 学号_____ 时间_____

评价项目及标准		分值	小组评议	教师评议
技术能力	1. 对基本概念的熟悉程度	10		
	2. 掌握主界面控件	10		
	3. 掌握配置界面控件	10		
	4. 掌握控件配置文件	10		
	5. 掌握自定义控件的实现方法	10		
	6. 掌握自定义控件的测试方法	10		
执行能力	1. 出勤情况	5		
	2. 遵守纪律情况	5		
	3. 是否主动参与，有无提问记录	5		
	4. 有无职业意识	5		
社会能力	1. 能否有效沟通	5		
	2. 是否使用基本的文明礼貌用语	5		
	3. 能否与组员主动交流、积极合作	5		
	4. 能否自我学习、自我管理	5		
总分		100		
评定等级：				
评价意见		学习意见		

评定等级具体如下。

A：优，得分>90。B：好，90≥得分>80。C：一般，80≥得分>60。D：有待提高，得分≤60。

小结

本项目首先介绍了 DJI 无人机的 App 窗口功能组件，其次介绍了基于 Widget 控件实现自定义控件 UI 的方法，最后介绍了通过 DJI PSDK 的 App 窗口功能组件自定义控件，实现在 DJI Pilot UI 上获取负载设备的系统时间的方法。

项目 7

时间同步与精准定位

任务 1 时间同步

任务要求

本任务将通过 DJI PSDK 的时间同步功能，消除负载设备与无人机系统的时钟差，并同步负载设备与无人机系统的时间。

知识导入

7.1 时间同步功能

时间同步功能用于消除负载设备与无人机系统的时钟差，同步负载设备与无人机系统的时间。具有时间同步功能的负载设备能够方便用户顺利地使用日志排查无人机飞行过程中的各类故障、分析传感器采样的数据、获取精准的定位信息等。在多数情况下，负载不联网且无时钟源，因此日志记录时间与真实时间有较大的误差，通过时间同步功能可以设置负载设备上的本地时间。另外，在定位应用中也需要使用时间同步功能来获取负载设备的精准定位结果。

PSDK 通过 PPS 信号（周期性的脉冲）同步负载设备和具有 RTK 功能的无人机系统的时间。在使用时间同步功能前，需要通过移动端 App 确认无人机与 RTK 卫星间保持良好的通信状态，该移动端 App 可为 DJI 发布的 App，如 DJI Pilot，也可为基于 MSDK 开发的移动端 App。

时间同步过程如下。

（1）将负载设备安装到无人机的云台上，在无人机上电后，使用 PSDK 开发的负载设备将接收到无人机发送的 PPS 信号。

（2）当检测到 PPS 信号的上升沿时，负载设备需要记录本地时间。

（3）PSDK 的底层处理程序将获取与 PPS 信号同步的无人机系统的时间。需要注意的是，应使用硬件中断的形式实现 PPS 信号的响应，并确保 PPS 信号上升沿到达负载设备与负载设备记录本地时间之间的延迟低于 1ms。

（4）PSDK 的底层处理程序将计算负载设备上的本地时间与无人机系统时间的时钟差，实现负载设备时间与无人机系统时间的同步。

（5）负载设备通过 DjiTimeSync_TransferToAircraftTime 函数将负载设备上的本地时间转换为无人机系统的时间。

7.2 接口说明

时间同步相关功能的头文件为 dji_time_sync.h，该文件描述了时间同步功能中结构体和函数原型的关键信息与使用方法。主要函数说明如下。

1. 初始化时间同步模块

DjiTimeSync_Init 函数用于在阻塞模式下初始化时间同步模块。用户应在所有其他时间同步操作之前调用此函数，就像传输时间一样。该函数的最大执行时间略大于 2000ms。该函数必须在用户任务中被调用，而不在 main 函数中被调用，且在调度程序启动后被调用。

```
T_DjiReturnCode DjiTimeSync_Init(void);
```

2. 获取最新的时间戳

注册的 DjiTimeSync_RegGetNewestPpsTriggerTimeCallback 回调函数用于在检测到 PPS 信号上升沿时获取本地时间系统中的最新时间戳。

```
T_DjiReturnCode DjiTimeSync_RegGetNewestPpsTriggerTimeCallback(
                DjiGetNewestPpsTriggerLocalTimeUsCallback callback);
```

说明：参数 callback 是一个指向回调函数的指针。

3. 时间转换

DjiTimeSync_TransferToAircraftTime 函数将负载设备上的本地时间转换为无人机系统的时间。在使用本函数时，需要调用 DjiTimeSync_RegGetNewestPpsTriggerTimeCallback 函数注册用于获取当负载设备检测到 PPS 信号上升沿时，负载设备本地最新的时间戳的回调函数。

```
T_DjiReturnCode DjiTimeSync_TransferToAircraftTime(
                uint64_t localTimeUs,
                T_DjiTimeSyncAircraftTime *aircraftTime);
```

说明：参数 localTimeUs 是负载设备上的本地时间，单位为微秒；参数 aircraftTime 指向用于存储无人机系统时间的内存空间。

7.3 代码模板

在开发者实现时间同步功能时，不要在 main 函数中调用时间同步的接口，而需要创建用户线程来实现时间同步功能。在用户线程中调用时间同步的接口，启动调度器后，时间同步接口将正常运行。

1. 代码结构

下面是实现时间同步功能的代码框架，用户只需要编写用于实现时间同步功能的 time_sync.c 和 time_sync.h 即可。

```
├── 3rdparty              // 第三方库文件目录，省略了该目录下的内容
├── CMakeLists.txt        // cmake 构建项目的文件
├── dji_init.c            // 系统初始化源文件
├── dji_init.h            // 系统初始化头文件
├── time_sync             // 用户实现时间同步的目录
│   ├── time_sync.c       // 实现时间同步的源文件
│   └── time_sync.h       // 头文件
├── hal                   // hal 文件目录，省略了该目录下的内容
├── main.c                // 主函数文件
├── monitor               // 监控文件目录，省略了该目录下的内容
├── osal                  // osal 文件目录，省略了该目录下的内容
└── psdk_lib              // psdk 库文件目录，省略了该目录下的内容
```

2. 头文件模板

本任务的头文件名为 time_sync.h。该文件定义的内容可满足正常的应用开发需求，如果需要实现特殊的功能，则可以按需要修改。

头文件中的 DjiTest_TimeSyncStartService 函数可以在 main.c 中被调用，作为时间同步功能的服务入口函数。本项目中的头文件 time_sync.h 无须修改，这里不再列出。

3. 源文件模板

本任务的源文件名为 time_sync.c，以下是 time_sync.c 的模板文件，按照提示即可完成代码的编写。

```c
/* Includes ------------------------------------------------------------------*/
#include "time_sync.h"
#include "dji_time_sync.h"
#include "dji_logger.h"
#include "utils/util_misc.h"
#include "dji_platform.h"
```

```c
/* Private constants ---------------------------------------------------------*/
#define DJI_TEST_TIME_SYNC_TASK_FREQ            (1)
#define DJI_TEST_TIME_SYNC_TASK_STACK_SIZE      (1024)

/* Private types -------------------------------------------------------------*/

/* Private functions declaration ---------------------------------------------*/
static void *DjiTest_TimeSyncTask(void *arg);

/* Private variables ---------------------------------------------------------*/
static T_DjiTestTimeSyncHandler s_timeSyncHandler;
static T_DjiTaskHandle s_timeSyncThread;

/* Exported functions definition ---------------------------------------------*/
T_DjiReturnCode DjiTest_TimeSyncRegHandler(T_DjiTestTimeSyncHandler *timeSyncHandler)
{
    if (timeSyncHandler->PpsSignalResponseInit == NULL) {
        USER_LOG_ERROR("reg time sync handler PpsSignalResponseInit error");
        return DJI_ERROR_SYSTEM_MODULE_CODE_INVALID_PARAMETER;
    }

    if (timeSyncHandler->GetNewestPpsTriggerLocalTimeUs == NULL) {
        USER_LOG_ERROR("reg time sync handler GetNewestPpsTriggerLocalTimeUs error");
        return DJI_ERROR_SYSTEM_MODULE_CODE_INVALID_PARAMETER;
    }

    memcpy(&s_timeSyncHandler, timeSyncHandler, sizeof(T_DjiTestTimeSyncHandler));

    return DJI_ERROR_SYSTEM_MODULE_CODE_SUCCESS;
}

/* ----在服务入口函数 DjiTest_TimeSyncStartService 中完成以下功能----*/
/* step 1 调用 DjiTimeSync_Init 函数初始化时间同步功能模块--------------*/
/* step 2 获取树莓派中断处理信号----------------------------------------*/
/* step 3 获取当前 PPS 触发时间----------------------------------------*/
/* step 4 注册获取最新 PPS 触发时间的回调函数--------------------------*/
/* step 5 调用 osalHandler->TaskCreate 函数启动时间转化线程 DjiTest_TimeSyncTask------*/

/* ---------------------------------------------------------------------------*/
T_DjiReturnCode DjiTest_TimeSyncStartService(void)
{
    T_DjiReturnCode djiStat;
    T_DjiOsalHandler *osalHandler = DjiPlatform_GetOsalHandler();

    /* step 1 调用 DjiTimeSync_Init 函数初始化时间同步功能模块*/
```

```
    /* step 2 获取树莓派中断处理信号--------------------------*/
    /* step 3 获取当前PPS触发时间----------------------------*/
    /* step 4 注册获取最新PPS触发时间的回调函数---------------*/
    /* step 5 调用osalHandler->TaskCreate 函数启动时间转化线程 DjiTest_TimeSyncTask-
-----*/

    return DJI_ERROR_SYSTEM_MODULE_CODE_SUCCESS;
}
T_DjiReturnCode DjiTest_TimeSyncGetNewestPpsTriggerLocalTimeUs(uint64_t
*localTimeUs)
{
    if (s_timeSyncHandler.GetNewestPpsTriggerLocalTimeUs == NULL) {
        USER_LOG_ERROR("GetNewestPpsTriggerLocalTimeUs null error.");
        return DJI_ERROR_SYSTEM_MODULE_CODE_INVALID_PARAMETER;
    }
    return s_timeSyncHandler.GetNewestPpsTriggerLocalTimeUs(localTimeUs);
}


/* ----在时间转化线程 DjiTest_TimeSyncTask 中完成以下功能--------*/

/* step 1 获取当前负载设备上的本地时间 --------------------------*/
/* step 2 当前负载设备时间对应的无人机系统的时间-----------------*/

/* -----------------------------------------------------------*/
static void *DjiTest_TimeSyncTask(void *arg)
{
    T_DjiReturnCode djiStat;
    uint32_t currentTimeMs = 0;
    T_DjiTimeSyncAircraftTime aircraftTime = {0};
    T_DjiOsalHandler *osalHandler = DjiPlatform_GetOsalHandler();
    USER_UTIL_UNUSED(arg);

    while (1) {
        osalHandler->TaskSleepMs(1000 / DJI_TEST_TIME_SYNC_TASK_FREQ);

        /* step 1 获取当前负载设备上的本地时间 --------------------------*/
        /* step 2 当前负载设备时间对应的无人机系统的时间-----------------*/
        USER_LOG_DEBUG("current aircraft time
is %04d-%02d-%02d %02d:%02d:%02d %d.",aircraftTime.year, aircraftTime.month,
aircraftTime.day,aircraftTime.hour, aircraftTime.minute, aircraftTime.second,
aircraftTime.microsecond);
    }
}
```

➡ 任务设计

时间同步功能用于消除负载设备与无人机系统的时钟差,同步负载设备与无人机系统的时间。任务的具体设计要求如下。

(1)创建项目目录。

(2)复制 psdk_lib、hal、osal、monitor、3rdparty、main.c 和 CMakeLists.txt 到该项目目录下。

(3)本任务将实现时间同步功能,项目目录下的 time_sync 目录用于编写代码,以实现用户线程调用时间同步 API 接口。

(4)在 time_sync 目录下编写代码,实现时间同步功能。

① 初始化时间同步模块。

② 获取树莓派中断处理信号。

③ 获取当前 PPS 触发时间。

④ 注册获取最新 PPS 触发时间的回调函数。

⑤ 调用启动时间转化线程。

⑥ 获取当前负载设备上的本地时间。

⑦ 获取当前负载设备时间对应的无人机时间。

(5)使用 cmake 编译项目。

(6)执行项目。

➡ 任务实施

在多数情况下,当负载设备不联网且无时钟源时,日志记录时间与真实时间会有较大的误差,可以通过时间同步功能来设置负载设备上的本地时间。限于篇幅,这里只给出任务实施步骤和关键代码。

1. 模板文件

连接树莓派开发板,默认路径为当前登录用户的家目录。查看当前的家目录,该目录下有一个名为 pro_7_sync_source_template 的目录,此目录下保存了本任务的所有模板文件。

目录结构如下。

```
pi@raspberrypi:~ $ cd pro_7_ sync_source_template/
pi@raspberrypi:~/pro_7_sync_source_template $ tree
.
├── 3rdparty              // 第三方库文件目录,省略了该目录下的内容
├── CMakeLists.txt        // cmake 构建项目的文件
├── dji_init.c            // 系统初始化源文件
├── dji_init.h            // 系统初始化头文件
├── time_sync             // 用户实现时间同步功能的目录
│   ├── time_sync.c       // 实现时间同步的源文件
│   ├── time_sync.h       // 头文件
```

```
├── hal                      // hal 文件目录，省略了该目录下的内容
├── main.c                   // 主函数文件
├── monitor                  // 监控文件目录，省略了该目录下的内容
├── osal                     // osal 文件目录，省略了该目录下的内容
└── psdk_lib                 // psdk 库文件目录，省略了该目录下的内容
```

2. 复制模板目录

将 pro_7_ sync_source_template 目录下的所有文件和目录复制到 pro_7_sync 目录中。

```
pi@raspberrypi:~ $ cp -r pro_7_ sync_source_template pro_7_sync
```

3. 修改 dji_init.c

开发者需根据实际的 PSDK 用户参数填写。

```
#define USER_App_NAME              "RaspberryPi"
#define USER_App_ID                "126413"
#define USER_App_KEY               "0ea2eadb314d1a6e870975364e2bf4a"
#define USER_App_LICENSE            "HZ/GptycWQ........................"
#define USER_DEVELOPER_ACCOUNT     "abc@outlo**.com"
#define USER_BAUD_RATE             "460800"
```

4. 修改 CMakeLists.txt

在 CMakeLists.txt 中需要修改 4 处，分别是修改项目的工程名、添加时间同步的目录和文件、添加库文件，按照下文提示找到相应的位置进行修改即可。

（1）设置本项目的工程名为 pro_7_sync。

```
# ----------------------设置工程名----------------------------
project(pro_7_sync C)
# -----------------------------------------------------------
```

（2）添加目录、文件和库文件。

```
# step 1 在此处加入时间同步目录下的所有.c 文件: time_sync/*.c
file(GLOB_RECURSE MODULE_SAMPLE_SRC time_sync/*.c)

# step 2 在此处加入时间同步目录: ./time_sync
include_directories(./time_sync)

# step 3 在此处加入树莓派 wiringPi 库文件/usr/lib/libwiringPi.so
link_libraries(/usr/lib/libwiringPi.so)
# -----------------------------------------------------------
```

修改后保存并退出。

5. 修改 main.c

（1）添加头文件。

```
#include <wiringPi.h>
#include "time_sync/time_sync.h"
```

（2）添加函数。

在 main 函数外部添加以下函数。

```c
static uint64_t s_ppsNewestTriggerLocalTimeUs;

/* step 1 定义上升沿信号检测到的处理函数 TimeSyncRTK_ppsIrpHandle--------*/
static void TimeSyncRTK_ppsIrpHandle(void)
{
    struct timeval tv;
    if(gettimeofday(&tv, NULL) == 0){              // 获取当前系统时间(精确到微秒)
        s_ppsNewestTriggerLocalTimeUs = (uint64_t)tv.tv_sec*1000000 + (uint64_t)tv.tv_usec;
        USER_LOG_INFO(" === get localtime: %u.%u us", tv.tv_sec, tv.tv_usec);
    }else{
        s_ppsNewestTriggerLocalTimeUs = 0;
    }
}
/* ----------------------------------------------------------------*/

/* step 2 定义树莓派边沿检测函数 TimeSyncRTK_setupPiIrq ------------- */
static T_DjiReturnCode TimeSyncRTK_setupPiIrq()
{
    /* step 2.1 初始化 wiringPi 功能 ----------------------- */
    if(-1 == wiringPiSetup()){
        USER_LOG_ERROR("wiringPi setup error");
        return DJI_RETURN_CODE_ERR_UNKNOWN;
    }
    /* ---------------------------------------------------*/

    /* step 2.2 定义 wiringPi 编码引脚 4 为输入模式，并设置为下拉模式 ---- */
    const int ppsPin = 4;                   // wiringPi 编码引脚

    pinMode(ppsPin, INPUT);                 // 设置输入模式
    pullUpDnControl(ppsPin, PUD_DOWN);      // 设置下拉模式
    /* ---------------------------------------------------*/

    /* step 2.3 设置 ppsPin 引脚输入信号的上升沿检测的回调处理函数--*/
    if( wiringPiISR(ppsPin, INT_EDGE_RISING, TimeSyncRTK_ppsIrpHandle) < 0){
        USER_LOG_ERROR("wiringPi interupt register error");
        return DJI_RETURN_CODE_ERR_UNKNOWN;
    }
    /* ---------------------------------------------------*/
    return DJI_RETURN_CODE_OK;
}

/* step 3 配置树莓派中断处理功能 TimeSyncRTK_PpsSignalResponseInit 函数---*/
T_DjiReturnCode TimeSyncRTK_PpsSignalResponseInit(void)
{
    if (TimeSyncRTK_setupPiIrq() != DJI_RETURN_CODE_OK){
```

```
        return DJI_RETURN_CODE_ERR_UNKNOWN;
    }
    USER_LOG_INFO("pps singnal response init success.");

    return DJI_ERROR_SYSTEM_MODULE_CODE_SUCCESS;
}
/* ----------------------------------------------------------------*/

/* step 4 定义 TimeSyncRTK_GetNewestPpsTriggerLocalTimeUs 函数，获取当前 PPS 触发时
间-----*/
T_DjiReturnCode TimeSyncRTK_GetNewestPpsTriggerLocalTimeUs(uint64_t *localTimeUs)
{
    if (localTimeUs == NULL) {
        USER_LOG_ERROR("input pointer is null.");
        return DJI_ERROR_SYSTEM_MODULE_CODE_INVALID_PARAMETER;
    }
    if (s_ppsNewestTriggerLocalTimeUs == 0) {
        USER_LOG_WARN("pps have not been triggered.");
        return DJI_ERROR_SYSTEM_MODULE_CODE_BUSY;
    }
    *localTimeUs = s_ppsNewestTriggerLocalTimeUs;    // 获取当前 PPS 触发时间
/* ---------------------------------------------------------------------*/
    return DJI_ERROR_SYSTEM_MODULE_CODE_SUCCESS;
}
```

（3）修改 main.c。

在 main 函数内部完成下面的功能。

```
int main(int argc, char **argv)
{
    /*------在此处完成时间同步模块的初始化-----------------------------------*/

    /* step 1 将树莓派中断处理函数和获取当前 PPS 触发时间函数赋给在 time_sync.h 中定义的结
构体 T_DjiTestTimeSyncHandler----*/
    T_DjiTestTimeSyncHandler timeSyncHandler = {
        .PpsSignalResponseInit = TimeSyncRTK_PpsSignalResponseInit,
        .GetNewestPpsTriggerLocalTimeUs = TimeSyncRTK_GetNewestPpsTriggerLocalTimeUs
    };
    DjiTest_TimeSyncRegHandler(&timeSyncHandler);
    /* ---------------------------------------------------------------------*/

    /* step 2 调用时间同步服务入口函数 DjiTest_TimeSyncStartService 完成初始化-----*/
    returnCode = DjiTest_TimeSyncStartService();
    if (returnCode != DJI_ERROR_SYSTEM_MODULE_CODE_SUCCESS) {
        USER_LOG_ERROR("power management init error");
    }
```

```
        /* ------------------------------------------------------------------*/
        return 0;
    }
```

6. 修改 time_sync.h

时间同步的头文件无须修改。

7. 修改 time_sync.c

（1）完成 DjiTest_TimeSyncStartService 函数。

根据模板文件，在 DjiTest_TimeSyncStartService 函数中添加相应的代码，完成后的代码如下。

```
T_DjiReturnCode DjiTest_TimeSyncStartService(void)
{
    T_DjiReturnCode djiStat;
    T_DjiOsalHandler *osalHandler = DjiPlatform_GetOsalHandler();

    /* step 1 调用 DjiTimeSync_Init 函数初始化时间同步模块*/
    djiStat = DjiTimeSync_Init();
    if (djiStat != DJI_ERROR_SYSTEM_MODULE_CODE_SUCCESS) {
        USER_LOG_ERROR("time synchronization module init error.");
        return djiStat;
    }
    /* ----------------------------------------------------------*/

    /* step 2 获取树莓派中断处理信号----------------------------*/
    if (s_timeSyncHandler.PpsSignalResponseInit == NULL) {
        USER_LOG_ERROR("time sync handler PpsSignalResponseInit interface is NULL error");
        return DJI_ERROR_SYSTEM_MODULE_CODE_UNKNOWN;
    }
    /* ----------------------------------------------------------*/

    /* step 3 获取当前 PPS 触发时间-----------------------------*/
    if (s_timeSyncHandler.GetNewestPpsTriggerLocalTimeUs == NULL) {
        USER_LOG_ERROR("time sync handler GetNewestPpsTriggerLocalTimeUs interface is NULL error");
        return DJI_ERROR_SYSTEM_MODULE_CODE_UNKNOWN;
    }
    /* ----------------------------------------------------------*/

    /* step 4 注册获取最新 PPS 触发时间的回调函数---------------*/
    djiStat = DjiTimeSync_RegGetNewestPpsTriggerTimeCallback
(s_timeSyncHandler.GetNewestPpsTriggerLocalTimeUs);
    if (djiStat != DJI_ERROR_SYSTEM_MODULE_CODE_SUCCESS) {
        USER_LOG_ERROR("register GetNewestPpsTriggerLocalTimeUsCallback error.");
```

```c
        return djiStat;
    }
    /* --------------------------------------------------------*/

    /* step 5 调用 osalHandler->TaskCreate 函数启动时间转化线程 DjiTest_TimeSyncTask-
-----*/
    if (osalHandler->TaskCreate("user_time_sync_task", DjiTest_TimeSyncTask,
                                DJI_TEST_TIME_SYNC_TASK_STACK_SIZE, NULL,
&s_timeSyncThread) !=
        DJI_ERROR_SYSTEM_MODULE_CODE_SUCCESS) {
        USER_LOG_ERROR("user time sync task create error.");
        return DJI_ERROR_SYSTEM_MODULE_CODE_UNKNOWN;
    }
    /* --------------------------------------------------------*/
}
```

（2）完成时间转化线程。

```c
/* ----在时间转化线程 DjiTest_TimeSyncTask 中完成以下功能-------*/

static void *DjiTest_TimeSyncTask(void *arg)
{
    while (1) {
        osalHandler->TaskSleepMs(1000 / DJI_TEST_TIME_SYNC_TASK_FREQ);

        /* step 1 获取当前负载设备上的本地时间 --------------------------------*/
        djiStat = osalHandler->GetTimeMs(&currentTimeMs);
        if (djiStat != DJI_ERROR_SYSTEM_MODULE_CODE_SUCCESS) {
            USER_LOG_ERROR("get current time error: 0x%08llX.", djiStat);
            continue;
        }
        /* ----------------------------------------------------------*/

        /* step 2 获取当前负载设备时间对应的无人机系统的时间----------------------*/
        djiStat = DjiTimeSync_TransferToAircraftTime(currentTimeMs * 1000,
&aircraftTime);
        if (djiStat != DJI_ERROR_SYSTEM_MODULE_CODE_SUCCESS) {
            USER_LOG_ERROR("transfer to aircraft time error: 0x%08llX.", djiStat);
            continue;
        }

        USER_LOG_DEBUG("current aircraft time is %04d-%02d-%02d %02d:%02d:%02d %d.",
                       aircraftTime.year, aircraftTime.month, aircraftTime.day,
                       aircraftTime.hour, aircraftTime.minute,
aircraftTime.second, aircraftTime.microsecond);
    }
    /* ----------------------------------------------------------*/
}
```

8. 编译并执行项目

```
# 在项目 pro_7_sync 目录下，创建 build 目录，并进入 build 目录
pi@raspberrypi:~/pro_7_sync$ mkdir build && cd build

# 使用 cmake 创建项目编译环境
pi@raspberrypi:~/pro_7_sync/build $ cmake ..

# make 编译
pi@raspberrypi:~/pro_7_sync/build $ make

# 执行文件位于 build/bin 目录下
pi@raspberrypi:~/pro_7_sync/build $cd bin/

# 使用 sudo 执行项目的执行文件
pi@raspberrypi:~/pro_7_sync/build/bin $ sudo ./pro_7_sync
```

任务评价

任务过程评价表

任务实施人姓名_____ 学号_____ 时间_____

	评价项目及标准	分值	小组评议	教师评议
技术能力	1. 对基本概念的熟悉程度	10		
	2. 掌握时间同步功能 PSDK 接口	10		
	3. 掌握时间同步功能的代码框架	10		
	4. 掌握时间同步模块初始化功能	10		
	5. 掌握时间同步功能注册回调函数的方法	10		
	6. 掌握时间同步功能中时间转化的方法	10		
执行能力	1. 出勤情况	5		
	2. 遵守纪律情况	5		
	3. 是否主动参与，有无提问记录	5		
	4. 有无职业意识	5		
社会能力	1. 能否有效沟通	5		
	2. 是否使用基本的文明礼貌用语	5		
	3. 能否与组员主动交流、积极合作	5		
	4. 能否自我学习、自我管理	5		
总分		100		
评定等级：				
评价意见		学习意见		

评定等级具体如下。

A：优，得分>90。B：好，90≥得分>80。C：一般，80≥得分>60。D：有待提高，得分≤60。

小结

本任务首先简要介绍了 DJI 无人机时间同步功能的相关知识；其次介绍了与时间同步功能相关的 PSDK 接口；最后介绍了时间同步功能的代码框架。

任务 2　精准定位

任务要求

本任务在时间同步功能的基础上，在负载端获取特定事件发生时的无人机精确定位位置，实现无人机的精准定位。

知识导入

7.4　精准定位功能

无人机的精准定位需要开启 RTK 并使 RTK fix，通过负载设备端与无人机端的消息联动可以指定并获取无人机记录的特定位置信息。这个消息就是定位事件，负载端的定位事件支持自定义，可以是拍照、人为触发或其他任何可以标定本地时间的事件。无人机端的定位事件只有一个，就是无人机系统时间，所以精准定位必须使用时间同步功能。通过时间同步功能使本地事件触发的时间对标无人机系统时间，并请求该时间下的无人机定位位置。

1. 基本概念

- 目标点：实际获取定位信息的位置，如云台口中心点。
- 兴趣点：由用户任意指定的负载设备中某个器件的位置，如相机图像传感器的中心点，该目标点也可为兴趣点。
- 任务：多个连续的飞行动作集合被称为一个任务，如对某个区域执行一次测绘任务。根据实际使用需要，用户可创建多个任务。
- 定位事件：触发定位请求的事件，如果相机在曝光时触发定位请求，则"相机曝光"是一个定位事件；多个事件的集合被称为事件集合，使用 PSDK 开发的负载设备可同时请求多个定位事件发生时的位置信息，如相机协同曝光。

2. 卫星间的通信状态

在运行"精准定位"代码前，请使用 DJI Pilot 或基于 MSDK 开发的 App 查看无人机与 RTK 卫星间是否保持着良好的通信状态，确保负载设备可获取精准的定位结果。

3. 获取精准定位

在获取精准定位时，需使用时间同步功能将负载设备上的本地时间同步为无人机系统的时间，有关使用时间同步功能的详细说明参见项目 7 的任务 1。获取精准定位的流程如下。

（1）当定位事件发生时，负载设备需要记录本地时间（该时间为负载设备上的本地时间）。

（2）负载设备通过时间转换功能将负载设备上的时间转换为无人机系统的时间。

（3）负载设备使用定位事件发生时的无人机系统的时间请求位置。

定位事件发生时的无人机系统的时间应早于最新的 PPS 信号上升沿时间，且时间间隔需在 1s～2s 内。

在获取目标点的位置后，根据目标点的位置，使用目标点与无人机 RTK 主天线位置的偏移量、无人机的姿态、负载设备的结构等信息，能够计算兴趣点的位置。

7.5 接口说明

精准定位相关功能的头文件为 dji_positioning.h，该文件描述了精准定位功能中结构体和函数原型的关键信息与使用方法。主要函数说明如下。

1. 初始化精准定位模块

DjiPositioning_Init 函数用于在使用精准定位功能前，调用本函数初始化精准定位模块。该函数初始化的时间可能会超过 500ms；请勿在主函数中调用该函数。

```
T_DjiReturnCode DjiPositioning_Init(void);
```

2. 设置任务索引

DjiPositioning_SetTaskIndex 函数用于设置任务索引。一次任务可能涉及对一个区域做航测建图，默认任务索引为 0。

```
void DjiPositioning_SetTaskIndex(uint8_t index);
```

3. 获取精准定位信息

DjiPositioning_GetPositionInformationSync 函数用于获取精准定位信息及其他信息。

```
T_DjiReturnCode DjiPositioning_GetPositionInformationSync(
uint8_t eventCount,
T_DjiPositioningEventInfo *eventInfo,
                 T_DjiPositioningPositionInfo *positionInfo);
```

说明：参数 eventCount 是同时获取定位信息的事件数量（<5）；参数 eventInfo 是指向定位时间的信息；参数 positionInfo 是目标点的定位信息。

7.6 代码模板

在开发者实现精准定位功能时,不要在 main 函数中调用精准定位的接口,而需要创建用户线程来实现精准定位功能。在用户线程中调用精准定位的接口,启动调度器后,精准定位接口将正常运行。

1. 代码结构

下面是实现精准定位功能的代码框架,用户只需要编写用于实现精准定位功能的 positioning.c 和 positioning.h 即可。

```
├── 3rdparty              // 第三方库文件目录,省略了该目录下的内容
├── CMakeLists.txt        // cmake 构建项目的文件
├── dji_init.c            // 系统初始化源文件
├── dji_init.h            // 系统初始化头文件
├── time_sync             // 用户实现时间同步的目录
│   ├── time_sync.c       // 实现时间同步的源文件
│   └── time_sync.h       // 头文件
├── info_mgmt             // 用户实现信息管理的目录
│   ├── info_mgmt.c       // 实现信息管理的源文件
│   └── info_mgmt.h       // 头文件
├── positioning           // 用户实现精准定位的目录
│   ├── positioning.c     // 实现精准定位的源文件
│   └── positioning.h     // 头文件
├── hal                   // hal 文件目录,省略了该目录下的内容
├── main.c                // 主函数文件
├── monitor               // 监控文件目录,省略了该目录下的内容
├── osal                  // osal 文件目录,省略了该目录下的内容
└── psdk_lib              // psdk 库文件目录,省略了该目录下的内容
```

说明:精准定位功能必须依赖时间同步功能,同时,获取信息还需要使用信息管理功能,因此实现精准定位功能需要引入时间同步功能(time_sync)和信息管理功能(info_mgmt)。

2. 头文件模板

本任务的头文件名为 positioning.h。该文件中定义的内容可满足正常的应用开发需求,如果需要实现特殊的功能,则可以按需要修改。

头文件中的 DjiTest_PositioningStartService 函数可在 main.c 中被调用,并可作为精准定位功能的服务入口函数。本任务的头文件 positioning.h 无须修改,这里不再列出。

3. 源文件模板

本任务的源文件名为 positioning.c,以下是 positioning.c 的模板文件,按照提示即可完成代码的编写。

```c
#include "positioning.h"
#include "dji_positioning.h"
#include "dji_logger.h"
#include "utils/util_misc.h"
#include "dji_platform.h"

/* ----------在此处添加信息管理和时间同步的头文件------------------------*/
/* step 1 添加时间同步的头文件 time_sync/time_sync.h  ------*/
/* step 2 添加信息管理的头文件 info_mgmt/info_mgmt.h  ------*/

/* --------------------------------------------------------------------*/

/* Private constants --------------------------------------------------*/
#define POSITIONING_TASK_FREQ                       (1)
#define POSITIONING_TASK_STACK_SIZE                 (1024)

#define DJI_TEST_POSITIONING_EVENT_COUNT            (2)
#define DJI_TEST_TIME_INTERVAL_AMONG_EVENTS_US      (200000)

/* Private types ------------------------------------------------------*/

/* Private functions declaration --------------------------------------*/
static void *DjiTest_PositioningTask(void *arg);

/* Private variables --------------------------------------------------*/
static T_DjiTaskHandle s_userPositioningThread;
static int32_t s_eventIndex = 0;

/* ---在精准定位服务入口函数 DjiTest_PositioningStartService 中完成以下功能-*/
/* step 1 调用 DjiPositioning_Init 函数初始化精准定位模块--------------*/
/* step 2 调用 DjiPositioning_SetTaskIndex 函数设置初始事件索引值为 0 -----*/
/* step 3 调用 osalHandler->TaskCreate 函数创建用户线程 DjiTest_PositioningTask--*/

T_DjiReturnCode DjiTest_PositioningStartService(void)
{
    T_DjiReturnCode djiStat;
    T_DjiOsalHandler *osalHandler = DjiPlatform_GetOsalHandler();

    /* step 1 调用 DjiPositioning_Init 函数初始化精准定位模块---------*/
    /* step 2 调用 DjiPositioning_SetTaskIndex 函数设置初始事件索引值为 0 --*/
    /* step 3 调用 osalHandler->TaskCreate 函数创建用户线程 DjiTest_PositioningTask-*/

    return DJI_ERROR_SYSTEM_MODULE_CODE_SUCCESS;
}

#ifndef __CC_ARM
```

```c
#pragma GCC diagnostic push
#pragma GCC diagnostic ignored "-Wmissing-noreturn"
#pragma GCC diagnostic ignored "-Wreturn-type"
#endif

/* -----在用户线程 DjiTest_PositioningTask 中完成以下功能------------*/
/* step 1 调用 DjiTest_FcSubscriptionGetTotalSatelliteNumber 函数获取卫星数-*/
/* step 2 调用 DjiTest_TimeSyncGetNewestPpsTriggerLocalTimeUs 函数通过 PPS 脉冲获取
本地时间------------*/
/* step 3 调用 DjiPositioning_GetPositionInformationSync 函数，通过指定的事件信息获
取无人机的定位记录-----*/

static void *DjiTest_PositioningTask(void *arg)
{
    int32_t i = 0;
    T_DjiReturnCode djiStat;
    uint64_t ppsNewestTriggerTimeUs = 0;
    T_DjiPositioningEventInfo eventInfo[DJI_TEST_POSITIONING_EVENT_COUNT] = {0};
    T_DjiPositioningPositionInfo positionInfo[DJI_TEST_POSITIONING_EVENT_COUNT]
= {0};
    T_DjiTimeSyncAircraftTime aircraftTime = {0};
    T_DjiOsalHandler *osalHandler = DjiPlatform_GetOsalHandler();
    uint8_t totalSatelliteNumber = 0;

    USER_UTIL_UNUSED(arg);

    while (1) {
        osalHandler->TaskSleepMs(1000 / POSITIONING_TASK_FREQ);

        /* step 1 调用 DjiTest_FcSubscriptionGetTotalSatelliteNumber 函数获取卫星
数----------------------*/
        /* step 2 调用 DjiTest_TimeSyncGetNewestPpsTriggerLocalTimeUs 函数通过 PPS
脉冲获取本地时间-------------*/

        // 下面的这段循环代码使用 PPS 的本地时间制作了一个时间标志，在转换成无人机系统时间时
先减去 1s，表示事件发生在最新 PPS 脉冲时间之前 1s
        for (i = 0; i < DJI_TEST_POSITIONING_EVENT_COUNT; ++i) {
            eventInfo[i].eventSetIndex = s_eventIndex;
            eventInfo[i].targetPointIndex = i;

            djiStat = DjiTimeSync_TransferToAircraftTime(
                ppsNewestTriggerTimeUs - 1000000 - i *
DJI_TEST_TIME_INTERVAL_AMONG_EVENTS_US, &aircraftTime);
            if (djiStat != DJI_ERROR_SYSTEM_MODULE_CODE_SUCCESS) {
                USER_LOG_ERROR("transfer to aircraft time error: 0x%08llX.",
djiStat);
```

```
                    continue;
                }

                eventInfo[i].eventTime = aircraftTime;
            }
            /* step 3 调用 DjiPositioning_GetPositionInformationSync 函数，通过指定的事
件信息来获取无人机的定位记录---------*/

            USER_LOG_DEBUG("request position of target points success.");
            USER_LOG_DEBUG("detail position information:");
            for (i = 0; i < DJI_TEST_POSITIONING_EVENT_COUNT; ++i) {
                USER_LOG_DEBUG("position solution property: %d.",
positionInfo[i].positionSolutionProperty);
                USER_LOG_DEBUG("pitchAttitudeAngle: %d\trollAttitudeAngle: %d\tyaw
AttitudeAngle: %d",positionInfo[i].uavAttitude.pitch,
positionInfo[i].uavAttitude.roll,positionInfo[i].uavAttitude.yaw);
                USER_LOG_DEBUG("northPositionOffset: %d\tearthPositionOffset: %d\t
downPositionOffset: %d",positionInfo[i].offsetBetweenMainAntennaAndTargetPoint.x,
positionInfo[i].offsetBetweenMainAntennaAndTargetPoint.y,positionInfo[i].offsetBe
tweenMainAntennaAndTargetPoint.z);
                USER_LOG_DEBUG("longitude: %.8f\tlatitude: %.8f\theight: %.8f",pos
itionInfo[i].targetPointPosition.longitude,positionInfo[i].targetPointPosition.la
titude,positionInfo[i].targetPointPosition.height);
                USER_LOG_DEBUG(
                    "longStandardDeviation: %.8f\tlatStandardDeviation: %.8f\thgtS
tandardDeviation: %.8f",positionInfo[i].targetPointPositionStandardDeviation.long
itude,positionInfo[i].targetPointPositionStandardDeviation.latitude,positionInfo[
i].targetPointPositionStandardDeviation.height);
            }

            s_eventIndex++;
        }
    }
```

➡ 任务设计

无人机的精准定位是负载设备端与无人机端的消息联动，通过精准定位可以指定获取无人机记录的特定位置信息。精准定位必须使用时间同步功能，通过时间同步功能使本地事件触发的时间对标无人机系统时间，并请求该时间下的无人机定位位置。

任务的具体设计要求如下。

（1）从精准定位 pro_7_positioning_source_template 项目模板文件中复制一个新项目 pro_7_positioning。

（2）本任务将实现精准定位功能，项目目录下的 positioning 目录用于编写代码，以实现用户线程调用精准定位 API 接口。

（3）在 positioning 目录下编写代码，实现精准定位功能。

① 初始化精准定位模块。

② 设置任务 ID。

③ 获取本地时间。

④ 时间转换。

⑤ 获取精准的定位信息。

（4）使用 cmake 编译项目。

（5）执行项目。

任务实施

限于篇幅，这里只给出任务实施步骤和关键代码。

1. 模板文件

连接树莓派开发板，默认路径为当前登录用户的家目录。查看当前的家目录，该目录下有一个名为 pro_7_positioning_source_template 的目录，此目录下保存了本任务的所有模板文件。

目录结构如下。

```
pi@raspberrypi:~ $ cd pro_7_ positioning_source_template/
pi@raspberrypi:~/pro_7_ positioning_source_template $ tree
.
├── 3rdparty                  // 第三方库文件目录，省略了该目录下的内容
├── CMakeLists.txt            // cmake 构建项目的文件
├── dji_init.c                // 系统初始化源文件
├── dji_init.h                // 系统初始化头文件
├── time_sync                 // 用户实现时间同步的目录
│   ├── time_sync.c           // 实现时间同步的源文件
│   └── time_sync.h           // 头文件
├── info_mgmt                 // 用户实现信息管理的目录
│   ├── info_mgmt.c           // 实现信息管理的源文件
│   └── info_mgmt.h           // 头文件
├── positioning               // 用户实现精准定位的目录
│   ├── positioning.c         // 实现精准定位的源文件
│   └── positioning.h         // 头文件
├── hal                       // hal 文件目录，省略了该目录下的内容
├── main.c                    // 主函数文件
├── monitor                   // 监控文件目录，省略了该目录下的内容
├── osal                      // osal 文件目录，省略了该目录下的内容
└── psdk_lib                  // psdk 库文件目录，省略了该目录下的内容
```

2. 复制模板目录

将 pro_7_positioning_source_template 目录下的所有文件和目录复制到 pro_7_positioning 目录中。

```
pi@raspberrypi:~ $ cp -r pro_7_ positioning_source_template pro_7_ positioning
```

3. 修改 dji_init.c

开发者需根据实际的 PSDK 用户参数填写。

```
#define USER_App_NAME              "RaspberryPi"
#define USER_App_ID                "126413"
#define USER_App_KEY               "0ea2eadb314d1a6e870975364e2bf4a"
#define USER_App_LICENSE           "HZ/GptycWQ........................"
#define USER_DEVELOPER_ACCOUNT     "abc@outlo**.com"
#define USER_BAUD_RATE             "460800"
```

4. 修改 CMakeLists.txt

在 CMakeLists.txt 中需要修改 8 处，分别是修改项目的工程名，添加时间同步、信息管理、精准定位的目录和文件，以及添加库文件，按照下文提示找到相应的位置进行修改即可。

（1）设置本项目的工程名为 pro_7_positioning。

```
# --------------------设置工程名--------------------------------
project(pro_7_positioning C)
# ------------------------------------------------------------
```

（2）添加目录、文件和库文件。

```
# step 1 在此处加入时间同步目录下的所有.c 文件: time_sync/*.c
file(GLOB_RECURSE MODULE_SAMPLE_SRC time_sync/*.c)

# step 2 在此处加入时间同步目录: ./time_sync
include_directories(./time_sync)

# step 3 在此处加入树莓派 wiringPi 库文件/usr/lib/libwiringPi.so
link_libraries(/usr/lib/libwiringPi.so)

# step 4 在此处加入信息管理目录下的所有.c 文件: info_mgmt/*.c
file(GLOB_RECURSE MODULE_SAMPLE_SRC_SUBSCRIPTION info_mgmt/*.c)

# step 5 在此处加入信息管理目录: ./info_mgmt
include_directories(./info_mgmt)

# step 6 在此处加入精准定位目录下的所有.c 文件: positioning/*.c
file(GLOB_RECURSE MODULE_SAMPLE_SRC_POSITION positioning/*.c)

# step 7 在此处加入精准定位目录: ./positioning
include_directories(./positioning)
# ------------------------------------------------------------
```

5. 修改 main.c

（1）在 main.c 中添加头文件。

```c
#include <wiringPi.h>
#include "time_sync/time_sync.h"
#include "positioning/positioning.h"
```

（2）添加函数。

在 main 函数外部添加以下函数。

```c
static uint64_t s_ppsNewestTriggerLocalTimeUs;

/* step 1 定义上升沿信号检测到的处理函数 TimeSyncRTK_ppsIrpHandle--------*/
static void TimeSyncRTK_ppsIrpHandle(void)
{
    struct timeval tv;
    if(gettimeofday(&tv, NULL) == 0){           // 获取当前系统时间(精确到微秒)
        s_ppsNewestTriggerLocalTimeUs = (uint64_t)tv.tv_sec*1000000 + (uint64_t)tv.tv_usec;
        USER_LOG_INFO(" === get localtime: %u.%u us", tv.tv_sec, tv.tv_usec);
    }else{
        s_ppsNewestTriggerLocalTimeUs = 0;
    }
}
/* -----------------------------------------------------------------*/


/* step 2 定义树莓派边沿检测函数 TimeSyncRTK_setupPiIrq ------------- */
static T_DjiReturnCode TimeSyncRTK_setupPiIrq()
{
    /* step 2.1 初始化 wiringPi 功能 ------------------------ */
    if(-1 == wiringPiSetup()){
        USER_LOG_ERROR("wiringPi setup error");
        return DJI_RETURN_CODE_ERR_UNKNOWN;
    }
    /* ----------------------------------------------------*/

    /* step 2.2 定义 wiringPi 编码引脚 4 为输入模式，并设置为下拉模式 --- */
    const int ppsPin = 4;                  // wiringPi 编码引脚

    pinMode(ppsPin, INPUT);                // 设置输入模式
    pullUpDnControl(ppsPin, PUD_DOWN);     // 设置下拉模式
    /* ----------------------------------------------------*/

    /* step 2.3 设置 ppsPin 引脚输入信号的上升沿检测的回调处理函数--*/
    if( wiringPiISR(ppsPin, INT_EDGE_RISING, TimeSyncRTK_ppsIrpHandle) < 0){
        USER_LOG_ERROR("wiringPi interupt register error");
        return DJI_RETURN_CODE_ERR_UNKNOWN;
    }
```

```
    /* ----------------------------------------------------*/
}

/* step 3 配置树莓派中断处理功能 TimeSyncRTK_PpsSignalResponseInit 函数---*/
T_DjiReturnCode TimeSyncRTK_PpsSignalResponseInit(void)
{
    if (TimeSyncRTK_setupPiIrq() != DJI_RETURN_CODE_OK){
        return DJI_RETURN_CODE_ERR_UNKNOWN;
    }
    USER_LOG_INFO("pps singnal response init success.");

    return DJI_ERROR_SYSTEM_MODULE_CODE_SUCCESS;
}
/* ----------------------------------------------------------------*/

/* step 4 定义 TimeSyncRTK_GetNewestPpsTriggerLocalTimeUs 函数获取当前 PPS 触发时间-
-----*/
    T_DjiReturnCode TimeSyncRTK_GetNewestPpsTriggerLocalTimeUs(uint64_t *localTimeUs)
{
    if (localTimeUs == NULL) {
        USER_LOG_ERROR("input pointer is null.");
        return DJI_ERROR_SYSTEM_MODULE_CODE_INVALID_PARAMETER;
    }
    if (s_ppsNewestTriggerLocalTimeUs == 0) {
        USER_LOG_WARN("pps have not been triggered.");
        return DJI_ERROR_SYSTEM_MODULE_CODE_BUSY;
    }
    *localTimeUs = s_ppsNewestTriggerLocalTimeUs; // 获取当前 PPS 触发时间
/* ----------------------------------------------------------------*/
}
```

（3）修改 main 函数。

在 main 函数内部完成下面的功能。

```
int main(int argc, char **argv)
{
    /*------在此处完成时间同步模块的初始化---------------------------*/

    /* step 1 将树莓派中断处理函数和获取当前 PPS 触发时间函数赋给在 time_sync.h 中定义的结
构体 T_DjiTestTimeSyncHandler----*/
        T_DjiTestTimeSyncHandler timeSyncHandler = {
            .PpsSignalResponseInit = TimeSyncRTK_PpsSignalResponseInit,
            .GetNewestPpsTriggerLocalTimeUs =
TimeSyncRTK_GetNewestPpsTriggerLocalTimeUs
        };
        DjiTest_TimeSyncRegHandler(&timeSyncHandler);
    /* ------------------------------------------------------------*/
```

```c
    /* step 2 调用时间同步服务入口函数 DjiTest_TimeSyncStartService 完成初始化-----*/
    returnCode = DjiTest_TimeSyncStartService();
    if (returnCode != DJI_ERROR_SYSTEM_MODULE_CODE_SUCCESS) {
        USER_LOG_ERROR("power management init error");
    }
    /* ----------------------------------------------------------------------*/

    /* step 3 调用精准定位服务入口函数 DjiTest_PositioningStartService 完成初始化-*/
    returnCode = DjiTest_PositioningStartService();
    if (returnCode != DJI_ERROR_SYSTEM_MODULE_CODE_SUCCESS) {
        USER_LOG_ERROR("positioning init error");
    }
    /* ----------------------------------------------------------------------*/
}
```

6. 修改 positioning.h

精准定位的头文件无须修改。

7. 修改 positioning.c

（1）添加头文件。

```c
#include "time_sync/time_sync.h"
#include <info_mgmt/info_mgmt.h>
```

（2）完成 DjiTest_PositioningStartService 函数。

根据模板文件，在 DjiTest_PositioningStartService 函数中添加相应的代码。

```c
T_DjiReturnCode DjiTest_PositioningStartService(void)
{
    T_DjiReturnCode djiStat;
    T_DjiOsalHandler *osalHandler = DjiPlatform_GetOsalHandler();

    /* step 1 调用 DjiPositioning_Init 函数初始化精准定位模块----------*/
    djiStat = DjiPositioning_Init();
    if (djiStat != DJI_ERROR_SYSTEM_MODULE_CODE_SUCCESS) {
        USER_LOG_ERROR("positioning module init error.");
        return djiStat;
    }
    /* --------------------------------------------------------------*/

    /* step 2 调用 DjiPositioning_SetTaskIndex 函数设置初始事件索引值为 0 ---*/
    DjiPositioning_SetTaskIndex(0);
    /* --------------------------------------------------------------*/

    /* step 3 调用 osalHandler->TaskCreate 函数创建用户线程 DjiTest_PositioningTask----*/
```

```
            if (osalHandler->TaskCreate("user_positioning_task", DjiTest_PositioningTask,
                                        POSITIONING_TASK_STACK_SIZE, NULL,
&s_userPositioningThread) !=
                DJI_ERROR_SYSTEM_MODULE_CODE_SUCCESS) {
                USER_LOG_ERROR("user positioning task create error.");
                return DJI_ERROR_SYSTEM_MODULE_CODE_UNKNOWN;
            }
        /* -------------------------------------------------------------*/


            return DJI_ERROR_SYSTEM_MODULE_CODE_SUCCESS;
        }
```

（3）完成用户线程代码的编写。

```
        static void *DjiTest_PositioningTask(void *arg)
        {
            while (1) {
                osalHandler->TaskSleepMs(1000 / POSITIONING_TASK_FREQ);

                /* step 1 调用 DjiTest_FcSubscriptionGetTotalSatelliteNumber 函数获取卫星
数---------------------------*/
                djiStat =
DjiTest_FcSubscriptionGetTotalSatelliteNumber(&totalSatelliteNumber);
                if (djiStat != DJI_ERROR_SYSTEM_MODULE_CODE_SUCCESS) {
                    USER_LOG_ERROR("get total satellite number error: 0x%08llX.",
djiStat);
                    continue;
                }
                /* -------------------------------------------------------------*/


                /* step 2 调用 DjiTest_TimeSyncGetNewestPpsTriggerLocalTimeUs 函数通过 PPS
脉冲获取本地时间--------------*/
                djiStat =
DjiTest_TimeSyncGetNewestPpsTriggerLocalTimeUs(&ppsNewestTriggerTimeUs);
                if (djiStat != DJI_ERROR_SYSTEM_MODULE_CODE_SUCCESS) {
                    USER_LOG_ERROR("get newest pps trigger time error: 0x%08llX.",
djiStat);
                    continue;
                }
                /* -------------------------------------------------------------*/


                // 下面的这段循环代码使用 PPS 的本地时间制作了一个时间标志，在转换成无人机系统时间时
先减去 1s，表示事件发生在最新 PPS 脉冲时间之前 1s
                for (i = 0; i < DJI_TEST_POSITIONING_EVENT_COUNT; ++i) {
                    eventInfo[i].eventSetIndex = s_eventIndex;
                    eventInfo[i].targetPointIndex = i;
```

```
                djiStat = DjiTimeSync_TransferToAircraftTime(
                    ppsNewestTriggerTimeUs - 1000000 - i *
DJI_TEST_TIME_INTERVAL_AMONG_EVENTS_US, &aircraftTime);
            if (djiStat != DJI_ERROR_SYSTEM_MODULE_CODE_SUCCESS) {
                USER_LOG_ERROR("transfer to aircraft time error: 0x%08llX.",
djiStat);
                continue;
            }

            eventInfo[i].eventTime = aircraftTime;
        }
        /* step 3 调用 DjiPositioning_GetPositionInformationSync 函数通过指定的事件
信息来获取无人机的定位记录---------*/
        djiStat =
DjiPositioning_GetPositionInformationSync(DJI_TEST_POSITIONING_EVENT_COUNT,
eventInfo, positionInfo);
        if (djiStat != DJI_ERROR_SYSTEM_MODULE_CODE_SUCCESS) {
            USER_LOG_ERROR("get position information error.");
            continue;
        }
        /* ----------------------------------------------------------------*/
    }
```

8. 编译并执行项目

```
# 在项目 pro_7_positioning 目录下，创建 build 目录，并进入 build 目录
pi@raspberrypi:~/pro_7_ positioning$ mkdir build && cd build

# 使用 cmake 创建项目编译环境
pi@raspberrypi:~/pro_7_ positioning/build $ cmake ..

# make 编译
pi@raspberrypi:~/pro_7_ positioning/build $ make

# 执行文件位于 build/bin 目录下
pi@raspberrypi:~/pro_7_ positioning/build $cd bin/

# 使用 sudo 执行项目的执行文件
pi@raspberrypi:~/pro_7_ positioning/build/bin $ sudo ./pro_7_ positioning
```

任务评价

任务过程评价表

任务实施人姓名_____ 学号_____ 时间_____

评价项目及标准		分值	小组评议	教师评议
技术能力	1. 对基本概念的熟悉程度	10		
	2. 掌握精准定位功能的 PSDK 接口	10		
	3. 掌握精准定位功能的代码框架	10		
	4. 掌握精准定位初始化功能	10		
	5. 掌握精准定位功能中时间同步功能的作用	10		
	6. 掌握使用精准定位功能获取位置的方法	10		
执行能力	1. 出勤情况	5		
	2. 遵守纪律情况	5		
	3. 是否主动参与，有无提问记录	5		
	4. 有无职业意识	5		
社会能力	1. 能否有效沟通	5		
	2. 是否使用基本的文明礼貌用语	5		
	3. 能否与组员主动交流、积极合作	5		
	4. 能否自我学习、自我管理	5		
总分		100		
评定等级：				
评价意见		学习意见		

评定等级具体如下。

A：优，得分>90。B：好，90≥得分>80。C：一般，80≥得分>60。D：有待提高，得分≤60。

小结

本任务首先简要介绍了 DJI 无人机精准定位功能的相关知识；其次介绍了与精准定位功能相关的 PSDK 接口；最后介绍了精准定位功能的代码框架。

项目 8　SDK 互联互通

本项目将学习 SDK 互联互通功能。SDK 互联互通功能的实现需要额外的环境和设备，所以本项目仅要求学生理解 SDK 互联互通功能，不做代码实现。

8.1　SDK 互联互通功能

8.1.1　概述

基于 OSDK 开发的应用程序能够与基于 MSDK 开发的移动端 App 和基于 PSDK 开发的负载设备相互通信，如用户通过基于 MSDK 开发的移动端 App 向无人机发送控制指令，控制负载设备和机载计算机执行指定的任务；基于 OSDK 开发的应用程序控制负载设备执行所需的动作，向基于 MSDK 开发的移动端 App 发送状态信息；基于 PSDK 开发的负载设备向基于 MSDK 开发的移动端 App 和基于 OSDK 开发的应用程序发送视频码流或文件等类型的数据。SDK 互联互通如图 8-1 所示。

使用 SDK 互联互通功能，开发者能够：
- 按需动态创建要使用的传输通道。
- 根据业务需求，创建通道并传输指定业务的数据。
- 全双工通信，数据收发同步。

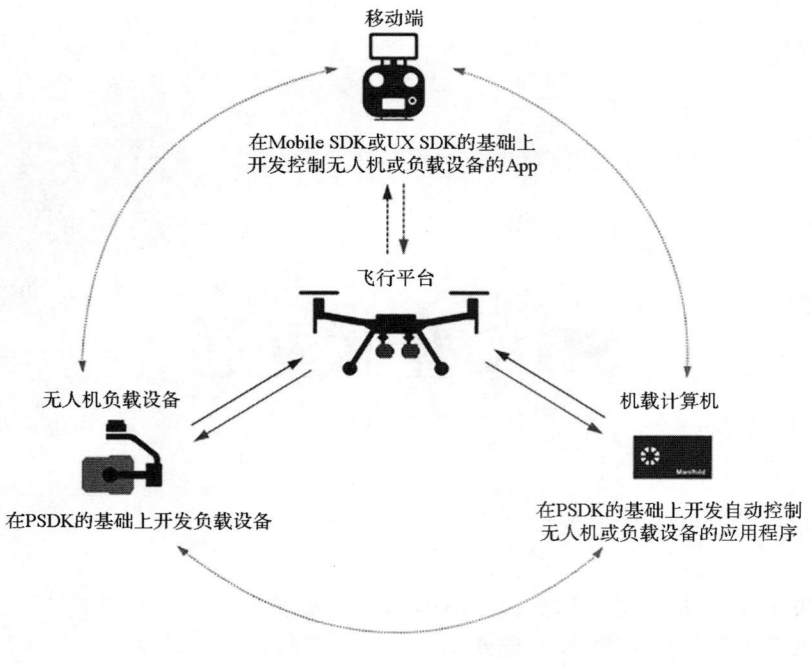

图 8-1　SDK 互联互通

8.1.2　基础概念

1．传输方式

（1）可靠传输：为确保基于不同 SDK 开发的应用程序和设备间能够实现可靠传输，DJI SDK 为开发者提供了可靠传输的方式，在该方式下，DJI SDK 内部采用了丢包重传、超时重发、错误检验等机制，确保不同 SDK 间收发的数据准确可靠。

- 数据可靠：以可靠传输的方式传输数据的双方，在发送和接收数据时需使用校验函数检验传输数据的正确性，同时 DJI SDK 使用加密算法加密所传输的数据，确保传输数据的安全性。
- 传输可靠：使用 DJI SDK 的数据传输功能可以拥有计时器和 ACK 机制，能够在数据传输超时后重发该数据，确保对端能够正常接收所发送的数据，防止数据意外丢失；此外，在以可靠传输的方式传输数据时，发送端将为所发送的数据编号，接收端根据数据编号能够重排接收到的乱序数据，确保数据传输的可靠性。

（2）不可靠传输：在以不可靠传输的方式传输数据时，使用不同 SDK 开发的应用程序和设备间能够以更快的速度传输数据，但无法保证可靠地传输数据。

2．对象指定

DJI SDK 的互联互通功能通过设备类型、设备槽位、通道 ID 能够准确指定无人机上所需通信的设备或模块。

（1）设备类型：为了方便开发者识别对端的身份和类型，更好地区分数据传输的对象，数据传输功能根据 DJI SDK 提供了 MSDK、OSDK 和 PSDK 三种设备类型。需要注意的是，在 SDK 互联互通功能中，只有基于 OSDK 开发的应用程序和基于 PSDK 开发的负载设备才可对外开放通道供基于 MSDK 开发的移动端 App 连接与使用，或者相互连接以传输数据。

（2）设备槽位：由于 DJI 无人机具有强大的扩展能力，开发者能够以多种拓展方式接入 3 台负载设备、双控及机载计算机等设备，因此可使用不同的槽位区分移动端 App、负载设备和机载计算机当前所处的位置。

（3）通道 ID：为了方便开发者选择和使用通信通道，区分同一个设备上的不同通道，DJI SDK 为互联互通功能提供了设置通道 ID 的功能，开发者在创建通道时可为创建的通道指定 ID。需要注意的是，只有基于 OSDK 开发的应用程序和基于 PSDK 开发的负载设备在使用数据传输功能时，才需要为创建的通道指定通道 ID。开发者可以根据实际的使用需求，在创建通道时，为所创建的通道指定通道 ID，最多支持设置 65535 个通道。

3. 通道管理

为了方便开发者快速开发出功能强大且可靠的应用程序和设备，无须关注 SDK 互联互通功能底层的工作逻辑，DJI SDK 为互联互通功能赋予了强大的通信管理能力。

（1）链路管理：管理链路的接入、关闭、销毁、重连及拥塞管理等。

（2）数据管理：数据转发、数据读写、流量控制、数据校验、乱序重排及丢包重传等。

需要注意的是，为了方便开发者使用 SDK 互联互通功能，与现有的接口进行区分，DJI MSDK 使用"Pipeline"表示通道管理功能。

4. 通道带宽

（1）MSDK 上行（向服务器端发送数据）可靠/非可靠传输最大速率一般为 24Kb/s～48Kb/s，带宽上限取决于实际的无线链路情况。

（2）MSDK 下行（从服务器端接收数据）可靠传输最大速率为 16Mb/s，非可靠传输最大速率为 20Mb/s。注意：MSDK 与服务器端（OSDK/PSDK）的通道带宽受限于无线链路，在远距离情况下，带宽上限会随着信号的变弱而降低，建议以实际测试情况为准。

（3）对于 OSDK 与 PSDK 间的数据传输，可靠传输最大速率为 24Mb/s，非可靠传输最大速率为 30Mb/s。

8.2 使用 SDK 互联互通功能

8.2.1 MSDK 端

当基于 MSDK 开发的移动端 App 使用 SDK 互联互通功能时，可根据用户的使用需求与指定的通道建立连接，实现数据的接收和发送；当无须使用 SDK 互联互通功能时，可断开已连接的通道。

基于 MSDK 开发的移动端 App 仅能作为客户端，通过通道 ID 与指定的通道建立连接。

1. 通道连接

通道连接时，请使用如下接口指定通道传输方式和需要连接的通道 ID，使该通道 ID 与指定的通道建立连接，并以阻塞回调的方式接收传输的数据。

- iOS。

```
-(void)connect:(uint16_t)Id pipelineType:(DJITransmissionControlType)transferType withCompletion:(void (^_Nullable)(DJIPipeline *_Nullable pipeline, NSError *_Nullable error))completion;
```

- Java。

```
void connect(int id, @NonNull TransmissionControlType transmissionType, @Nullable CommonCallbacks.CompletionCallback<PipelineError> callback);
```

2. 数据接收

通道建立连接后，基于 MSDK 开发的移动端 App 将使用如下接口接收对端发送的数据。

- iOS。

```
- (NSData *)readData:(uint32_t)readLength error:(NSError **)error;
```

- Java。

```
int readData(byte[] buff, int length);
```

3. 数据发送

为了实现良好的数据传输效果，建议每次传输的数据大小不超过 1KB。

- iOS。

```
- (int32_t)writeData:(NSData *)data error:(NSError **)error;
```

- Java。

```
int writeData(byte[] data);
```

4. 通道关闭

通信结束后，请使用如下接口断开已连接的通道。执行通道关闭的操作后，基于 MSDK 开发的移动端 App 将清除本地的缓存信息。通道关闭后，DJI SDK 将自动销毁已创建的通道，释放通道所占用的系统资源。

- iOS。

```
- (void)disconnect:(uint16_t)Id withCompletion:(DJICompletionBlock)completion;
```

- Java。

```
void disconnect(int id, @Nullable CommonCallbacks.CompletionCallback<PipelineError> callback);
```

8.2.2 OSDK 端

只有使用 Linux 平台的机载计算机才支持开发者基于 OSDK 开发互联互通功能。基于 OSDK 开发的应用程序既可以作为客户端，又可以作为服务器端。

- 当作为客户端时，用户可通过通道 ID 与指定的通道建立连接。
- 当作为服务器端时，基于 OSDK 开发的应用程序可根据用户的使用需求创建通道并指定通道 ID，仅当服务器端与客户端建立连接后，服务器端方可读写数据。

1. 初始化

如果基于 OSDK 开发的应用程序需要使用 SDK 互联互通功能,则需要先初始化 SDK 互联互通模块。

- 应用程序作为客户端。

当基于 OSDK 开发的应用程序作为客户端时，可将该机载计算机视为无人机的负载设备，因此，请使用 DJI PSDK 中的 PSDKManager 类初始化 SDK 互联互通模块。

```
ErrorCode::ErrorCodeType ret = vehicle->psdkManager->initPSDKModule(
    PAYLOAD_INDEX_0, "Main_psdk_device");
if (ret != ErrorCode::SysCommonErr::Success) {
  DERROR("Init PSDK module Main_psdk_device failed.");
  ErrorCode::printErrorCodeMsg(ret);
}
```

- 应用程序作为服务器端。

当基于 OSDK 开发的应用程序作为服务器端时，需要先初始化 Vehicle 类，并通过调用 vehicle->mopServer 中的 accept 接口阻塞指定的通道 ID。

```
#define TEST_MO_PIPELINE_ID 20

...
   if (vehicle->mopServer->accept((PipelineID)TEST_MO_PIPELINE_ID, UNRELIABLE,
MO_Pipeline) != MOP_PASSED) {
      DERROR("MOP Pipeline accept failed");
      return -1;
    } else {
      DSTATUS("Accept to mop pipeline id(%d) successfully", TEST_MO_PIPELINE_ID);
    }
...
```

2. 获取客户端对象的指针

仅当基于 OSDK 开发的应用程序作为客户端时，才需要获取使用 SDK 互联互通功能的客户端的指针，并通过该指针创建数据传输的通道或读写所需传输的数据。

```
MopClient *mopClient = NULL;
ret = vehicle->psdkManager->getMopClient(PAYLOAD_INDEX_0, mopClient);
if (ret != ErrorCode::SysCommonErr::Success) {
  DERROR("Get MOP client object for_psdk_device failed.");
  ErrorCode::printErrorCodeMsg(ret);
  return NULL;
}

if (!mopClient) {
  DERROR("Get MOP client object is a null value.");
  return NULL;
}
```

3. 通道连接

仅当基于 OSDK 开发的应用程序作为客户端时,才需要指定所要连接的通道 ID 和通道类型,并与指定的通道建立连接。

```
#define TEST_OP_PIPELINE_ID 15

MopPipeline *OP_Pipeline = NULL;
if ((mopClient->connect(TEST_OP_PIPELINE_ID, RELIABLE, OP_Pipeline)
   != MOP_PASSED) || (OP_Pipeline == NULL)) {
  DERROR("MOP Pipeline connect failed");
  return NULL;
} else {
  DSTATUS("Connect to mop pipeline id(%d) successfully", TEST_OP_PIPELINE_ID);
}
```

4. 数据接收

创建通道后,开发者可在该通道上以阻塞回调(同步回调)的方式接收对端传输的数据。

- 客户端接收数据。

当基于 OSDK 开发的应用程序作为客户端时,请使用如下接口接收服务器端发送的数据。

```
MopErrCode mopRet;
...
uint8_t recvBuf[1024];
MopPipeline::DataPackType readPack = {recvBuf, 1024};
mopRet = OP_Pipeline->recvData(readPack, &readPack.length);
...
```

- 服务器端接收数据。

当基于 OSDK 开发的应用程序作为服务器端时,请使用如下接口读取客户端发送的数据。

```
MopErrCode mopRet;
...
uint8_t recvBuf[1024];
MopPipeline::DataPackType readPack = {recvBuf, 1024};
mopRet = MO_Pipeline->recvData(readPack, &readPack.length);
...
```

5. 数据发送

创建通道后,开发者可以在该通道上向对端发送数据。

- 客户端发送数据。

当基于 OSDK 开发的应用程序作为客户端时,请使用如下接口向服务器端发送数据。

```
MopErrCode mopRet;
uint8_t sendBuf[1024];
...
MopPipeline::DataPackType reqPack = {sendBuf, 1024};
mopRet = OP_Pipeline->sendData(reqPack, &reqPack.length);
...
```

- 服务器端发送数据。

当基于 OSDK 开发的应用程序作为服务器端时,请使用如下接口向客户端发送数据。

```
uint8_t sendBuf[1024];
...
MopPipeline::DataPackType reqPack = {sendBuf, 1024};
mopRet = MO_Pipeline->sendData(reqPack, &reqPack.length);
...
```

6. 通道关闭

通信结束后,请使用如下接口断开与指定通道的连接,并释放通道占用的系统资源。

- 客户端关闭通道。

当基于 OSDK 开发的应用程序作为客户端时,请使用如下接口关闭已创建的通道。

```
if (mopClient->disconnect(TEST_OP_PIPELINE_ID) != MOP_PASSED) {
  DERROR("MOP Pipeline disconnect pipeline(%d) failed", TEST_OP_PIPELINE_ID);
} else {
  DSTATUS("Disconnect mop pipeline id(%d) successfully", TEST_OP_PIPELINE_ID);
}
```

- 服务器端关闭通道。

当基于 OSDK 开发的应用程序作为服务器端时,请使用如下接口关闭已创建的通道。

```
if (vehicle->mopServer->close(TEST_MO_PIPELINE_ID) != MOP_PASSED) {
  DERROR("MOP Pipeline disconnect pipeline(%d) failed", TEST_MO_PIPELINE_ID);
} else {
  DSTATUS("Disconnect mop pipeline id(%d) successfully", TEST_MO_PIPELINE_ID);
}
```

8.2.3 PSDK 端

只有使用 Linux 平台开发的负载设备才支持开发者基于 PSDK 开发互联互通功能。基于 PSDK 开发的负载设备只可作为服务器端,可根据用户的使用需求创建通道并指定通道 ID,仅当与客户端创建连接后,方可读写数据。

1. SDK 互联互通模块初始化

如果基于 PSDK 开发的负载设备需要使用 SDK 互联互通功能,则需要先初始化 SDK 互联互通模块。

```
T_DjiReturnCode DjiMopChannel_Init(void);
```

2. 创建通道

基于 PSDK 开发的负载设备根据用户指定的需求,创建相应的通道类型:可靠传输和不可靠传输。

```
T_DjiReturnCode DjiMopChannel_Create(T_DjiMopChannelHandle *channelHandle,
E_DjiMopChannelTransType transType);
```

3. 通道连接

基于 PSDK 开发的负载设备作为服务器端，在与对端建立连接时，需指定通道 ID，以供客户端绑定；为了方便同时与多个客户端建立连接，PSDK 提供了 outChannelHandle 句柄。

- 通道绑定：基于 PSDK 开发的负载设备通过指定的通道 ID 与客户端通信。

```
T_DjiReturnCode DjiMopChannel_Bind(T_DjiMopChannelHandle channelHandle,
                    uint16_t channelId);
```

- 接受连接：基于 PSDK 开发的负载设备通过如下接口接受对端发送的连接请求。

```
T_DjiReturnCode DjiMopChannel_Accept(
                    T_DjiMopChannelHandle channelHandle,
                    T_DjiMopChannelHandle *outChannelHandle);
```

说明：该接口为阻塞式的接口，当基于 PSDK 开发的负载设备作为服务器端时，为了能够同时与多个客户端建立连接，请在单独的线程中调用该接口。

4. 数据接收

创建通道后，开发者可在该通道上接收对端传输的数据。

```
T_DjiReturnCode DjiMopChannel_RecvData(
                    T_DjiMopChannelHandle channelHandle,
                        uint8_t *data,
                        uint32_t len,
                        uint32_t *realLen);
```

5. 数据发送

创建通道后，开发者可在该通道上向对端发送数据。

```
T_DjiReturnCode DjiMopChannel_SendData(
                    T_DjiMopChannelHandle channelHandle,
                        uint8_t *data,
                        uint32_t len,
                        uint32_t *realLen);
```

6. 通道关闭

通信结束后，请使用如下接口断开与指定通道的连接，并释放通道占用的系统资源。

- 关闭通道：调用如下接口关闭已创建的通道，关闭后，该通道将无法收发数据，但可与其他通道重新建立连接。

```
T_DjiReturnCode DjiMopChannel_Close(
                    T_DjiMopChannelHandle channelHandle);
```

- 销毁通道：调用如下接口销毁指定的通道。

```
T_DjiReturnCode DjiMopChannel_Destroy(
                    T_DjiMopChannelHandle channelHandle);
```

8.3 接口说明

SDK 互联互通相关功能的头文件为 dji_mop_channel.h,该文件描述了 SDK 互联互通功能中结构体和函数原型的关键信息与使用方法。说明：只有在 Linux 平台上开发的负载设备才支持开发者使用 SDK 互联互通功能。主要函数说明如下。

1. 通道初始化

```
T_DjiReturnCode DjiMopChannel_Init(void);
```

说明：在使用 MOP 通道传输信息前，请先使用本函数初始化 MOP 通道；在调用该函数前，请确认负载设备已接入网口。

2. 创建 MOP 通道

```
T_DjiReturnCode DjiMopChannel_Create(
            T_PsdkMopChannelHandle *channelHandle,
            E_DjiMopChannelTransType transType);
```

说明：参数 channelHandle 指向待创建的 MOP 通道句柄；参数 transType 是 MOP 通道的类型，详情请参见 E_DjiMopChannelTransType。MOP 通道创建成功后，请指定需要与已创建的通道建立连接的通道 ID，等待基于 MSDK 开发的移动端 App 或基于 OSDK 开发的应用程序连接该通道，成功接受其他设备的连接后，用户可在该通道上发送或接收数据；若无须使用 MOP 通道，则请关闭或销毁已创建的通道，并释放通道占用的系统资源。

3. 销毁已创建的 MOP 通道

```
T_DjiReturnCode DjiMopChannel_Destroy(
            T_PsdkMopChannelHandle channelHandle);
```

说明：参数 channelHandle 指向已创建的 MOP 通道。调用此函数后，开发者无法对此通道执行其他操作。

4. 信道绑定

```
T_DjiReturnCode DjiMopChannel_Bind(
            T_PsdkMopChannelHandle channelHandle,
                uint16_t channelId);
```

说明：参数 channelHandle 指向已创建的 MOP 通道；参数 channelId 指定所需要连接的通道 ID。成功创建 MOP 通道后，开发者可通过指定通道 ID 与已创建的 MOP 通道绑定，在调用函数 DjiMopChannel_Accept 后，可以与客户端设备建立连接。

5. 接受 MOP 通道连接

```
T_DjiReturnCode DjiMopChannel_Accept(
            T_PsdkMopChannelHandle channelHandle,
            T_PsdkMopChannelHandle *outChannelHandle);
```

说明：参数 channelHandle 指向已创建的 MOP 通道；参数 outChannelHandle 指向成功连接后的 MOP 输出通道。MOP 通过 channelHandle 监听通道接收在队列中等待连接的 MOP 通道的连接需求，创建一个新的已连接的通道，并通过 outChannelHandle 返回通道信息，负载设备作为服务器端允许多个客户端通过绑定通道 ID 与已创建的通道建立连接。

6. MOP 通道连接

```
T_DjiReturnCode DjiMopChannel_Connect(
                T_DjiMopChannelHandle channelHandle,
                E_DjiChannelAddress channelAddress,
                    uint16_t channelId);
```

说明：参数 channelHandle 指向已创建的 MOP 通道；参数 channelAddress 是通道地址；参数 channelId 是通道 ID。

7. 关闭已创建的 MOP 通道

```
T_DjiReturnCode DjiMopChannel_Close(
                T_PsdkMopChannelHandle channelHandle);
```

说明：参数 channelHandle 指向已创建的 MOP 通道。

8. 发送数据

```
T_DjiReturnCode DjiMopChannel_SendData(
                T_DjiMopChannelHandle channelHandle,
                            uint8_t *data,
                            uint32_t len,
                            uint32_t *realLen);
```

说明：参数 channelHandle 指向调用 DjiMopChannel_Accept 函数后的 MOP 句柄 outChannelHandle；参数 data 指向所需发送的数据；参数 len 是所需发送的数据的长度，单位为 Byte；参数 realLen 指向已发送数据的实际长度。

9. 接收数据

```
T_DjiReturnCode DjiMopChannel_RecvData(
                T_DjiMopChannelHandle channelHandle,
                            uint8_t *data,
                            uint32_t len,
                            uint32_t *realLen);
```

说明：参数 channelHandle 指向调用 DjiMopChannel_Accept 函数后的 MOP 句柄 outChannelHandle；参数 data 指向接收后存储的数据；参数 len 是所接收的数据的长度，单位为 Byte；参数 realLen 指向已接收数据的实际长度。

项目 9

视频流传输与回放下载

任务 1　视频流文件传输与回放下载

● 任务要求

本任务使用 PSDK 提供的视频流文件传输控制功能从负载设备上获取 H.264 格式的视频流文件，并在 DJI Pilot 遥控器上播放，同时使用回放下载功能从负载设备上下载 MP4 格式的视频文件，用于回放。

● 知识导入

9.1　概述

本任务涉及 PSDK 的多个功能，包括高速数据传输、视频流文件传输和回放下载等。

1. 高速数据传输

在数据传输功能中，根据通道带宽可将数据通道分为小数据传输通道和高速数据传输通道。其中，小数据传输通道通过串口通信，通常作为指令等数据的传输通道。高速数据传输通过网口链路，可以提供达 4Mb/s 的带宽通道。

显然，对于本项目中的视频流和下载流，适用于小数量传输的串口通道是无法满足的，需要使用高速数据传输通道。PSDK 是通过 UDP 实现高速数据传输通道的。在发送数据时可以通过 API 读取对端（无人机端）的 IP 地址和 PORT。

2. 视频流文件传输

当使用 PSDK 提供的视频流文件传输控制功能时,开发者需要先实现获取视频流文件码流的功能,按照 H.264 编码格式对视频流进行编码,并结合视频的帧率等参数,调用指定的接口发送视频流数据。用户通过使用 DJI Pilot 或基于 MSDK 开发的移动端 App,能够获取相机类负载设备上实时的视频流数据。

3. 回放下载

为了方便开发者快速开发相机类负载设备的回放下载功能,开发者需要先实现相机类负载设备的回放下载功能,再将回放下载功能的函数注册到指定的接口中,用户通过使用 DJI Pilot 或基于 MSDK 开发的移动端 App,即可获取相机类负载设备中的媒体文件。

9.2 数据传输

PSDK 支持用户为不同类型的数据设置高速数据传输通道带宽的占用比例,实现对高速数据传输通道带宽占用的精准控制。

为不同类型的数据设置高速数据传输通道带宽的占用比例后,基于 PSDK 开发的负载设备通过调用 DjiHighSpeedDataChannel_SetBandwidthProportion 函数,能够设置视频流等类型的数据占用高速数据传输通道带宽的比例。

在高速数据传输通道相关功能的头文件 dji_high_speed_data_channel.h 中定义了高速数据传输通道带宽比例的结构体 T_DjiDataChannelBandwidthProportionOfHighspeedChannel。

```
typedef struct {
    uint8_t dataStream;         //数据流的带宽比例,范围是0%~100%
    uint8_t videoStream;        //视频流的带宽比例,范围是0%~100%
    uint8_t downloadStream;     //下载流的带宽比例,范围是0%~100%
} T_DjiDataChannelBandwidthProportionOfHighspeedChannel;
```

说明:此结构体用于定义高速数据传输通道的每个流的带宽比例,带宽比例的总和必须为 100%。

同时,头文件中定义了函数 DjiHighSpeedDataChannel_SetBandwidthProportion,用于设置高速数据传输通道的数据流、视频流与下载流的带宽比例。

```
// 定义高速数据传输通道的每个流的带宽比例分别为10%、60%和30%
const T_DjiDataChannelBandwidthProportionOfHighspeedChannel
bandwidthProportion = {10, 60, 30};

// 设置高速数据传输通道的每个流的带宽比例
returnCode = DjiHighSpeedDataChannel_SetBandwidthProportion(bandwidthProportion);
if (returnCode != DJI_ERROR_SYSTEM_MODULE_CODE_SUCCESS) {
    USER_LOG_ERROR("Set data channel bandwidth width proportion error.");
    return DJI_ERROR_SYSTEM_MODULE_CODE_UNKNOWN;
}
```

9.3 配置网络参数

视频流文件传输与回放下载功能需要使用高速数据传输，PSDK 是通过网口链路，并使用 UDP 实现高速数据传输通道的。

1. 开启网络功能

在整个项目的初始化 dji_init.c 文件中，通过将 CONFIG_HARDWARE_CONNECTION 设置为 DJI_USE_UART_AND_NETWORK_DEVICE 同时开启串口和网络功能。

```
/*--------定义通信参数--------*/
#define DJI_USE_ONLY_UART                           (0)
#define DJI_USE_UART_AND_USB_BULK_DEVICE            (1)
#define DJI_USE_UART_AND_NETWORK_DEVICE             (2)
#define CONFIG_HARDWARE_CONNECTION  DJI_USE_UART_AND_NETWORK_DEVICE
```

网络功能开启后，开发者可根据实际情况选择以自动或手动的方式设置负载设备的网络参数。

2. 修改网络接口

PSDK 是通过网口链路进行视频流文件传输与回放下载的，在树莓派上使用 ifconfig 命令查看网络接口，如图 9-1 所示。

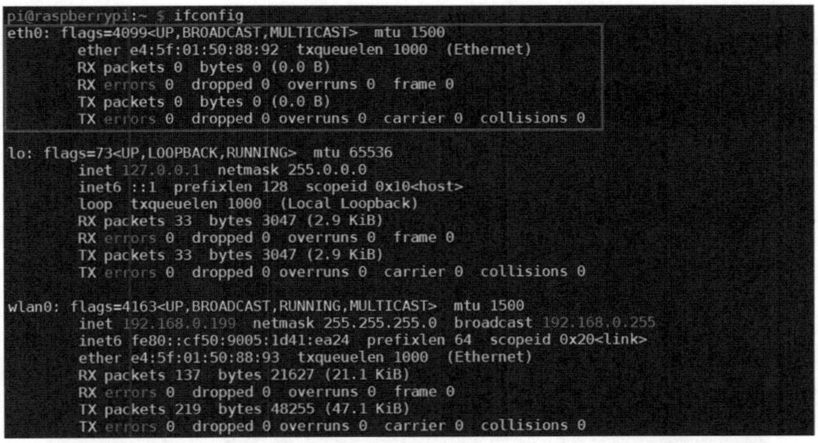

图 9-1 查看树莓派网络接口

从图 9-1 中可以看到，当前树莓派共有 3 个网络接口。其中，eth0 是物理网口，用于连接无人机扩展板；lo 是本地环回接口；wlan0 是无线网络，用于通过 SSH 登录树莓派。

为了让负载设备（树莓派）能使用高速数据传输与无人机进行视频流文件传输和回放下载，需要修改 hal/hal_network.h 中的 LINUX_NETWORK_DEV 为 eth0，如图 9-2 所示。

无人机应用技术开发（DJI Payload SDK）

```
/* Exported constants --------------------------------------*/
/** @attention  User can config network card name here, if your device is not MF2C/G, please comment below and add your
 *  NIC name micro define as #define "LINUX_NETWORK_DEV"  "your NIC name".
 */
#ifdef PLATFORM_ARCH_x86_64
#define LINUX_NETWORK_DEV           "enx000ec6688213"
#else
#define LINUX_NETWORK_DEV           "eth0"
#endif
```

图 9-2　修改树莓派网络接口配置

3. 以自动的方式设置网络参数（推荐）

为了方便开发者使用具有网络接口的开发平台实现更广泛的应用，PSDK 推荐开发者使用 DjiPlatform_RegHalNetworkHandler 接口注册设置负载设备网络参数的回调函数，这种方式能以自动的方式设置网络参数，否则负载设备仅能以手动的方式设置负载设备的网络参数。

当以自动的方式设置网络参数时，开发者需使用 DjiPlatform_RegHalNetworkHandler 接口注册设置负载设备网络参数的回调函数。

```
// 构造设置网络参数的回调函数。定义在 hal/hal_network.h 中
T_DjiReturnCode HalNetWork_Init(const char *ipAddr, const char *netMask,
T_DjiNetworkHandle *halObj)

// 注册设置负载设备网络参数的回调函数。在 dji_init.c 中调用该函数，并完成初始化
returnCode = DjiPlatform_RegHalNetworkHandler(&networkHandler);
if (returnCode != DJI_ERROR_SYSTEM_MODULE_CODE_SUCCESS) {
    printf("register hal network handler error");
    return DJI_ERROR_SYSTEM_MODULE_CODE_SYSTEM_ERROR;
```

注意事项：

① 树莓派与 PSDK 扩展板之间使用网线直接连接。

② 以自动的方式成功设置负载设备的网络参数后，负载设备默认的 IP 地址为 192.168.120.2/16。查看负载设备 IP 地址，如图 9-3 所示。无人机默认的 IP 地址为 192.168.110.1/16。负载设备 ping 无人机，如图 9-4 所示。

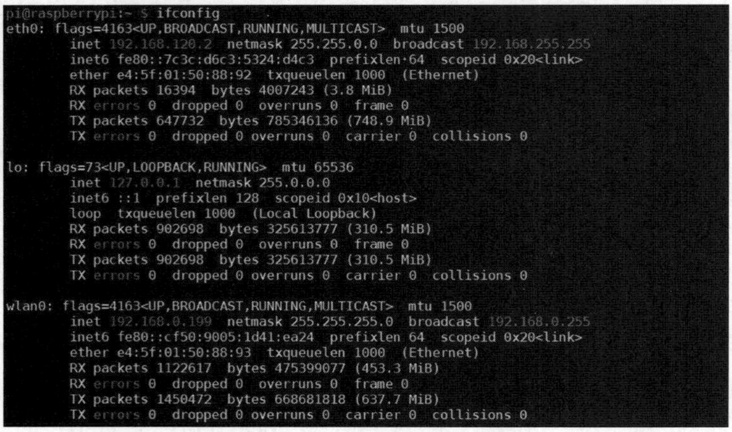

图 9-3　查看负载设备 IP 地址

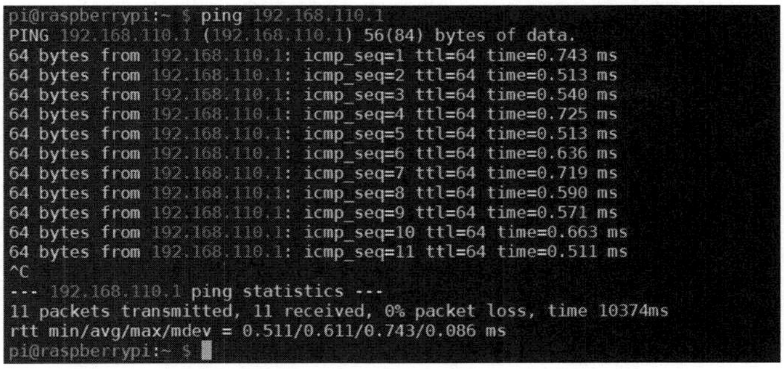

图 9-4　负载设备 ping 无人机

4. 以手动的方式设置网络参数

若不调用 DjiPlatform_RegHalNetworkHandler 接口注册设置负载设备网络参数的回调函数，则开发者需要手动设置网络参数，用户仅可使用视频流文件传输功能，其他涉及使用网络接口的功能将无法使用。例如，负载设备 IP 地址为 192.168.5.3/24，无人机 IP 地址为 192.168.5.1。IP 地址设置完成后，使用 ping 和 ifconfig 命令查看负载设备与无人机间的网络状态。

9.4　视频流文件传输

1. 相机模块初始化

视频流文件传输依赖相机模块，需要先初始化相机功能，才能正常使用视频流文件传输功能，同时需要初始化相机媒体流功能。在 main.c 中完成初始化。

```
// 初始化相机功能
returnCode = DjiTest_CameraEmuBaseStartService();
    if (returnCode != DJI_ERROR_SYSTEM_MODULE_CODE_SUCCESS) {
        USER_LOG_ERROR("camera emu common init error");
}

// 初始化相机媒体流功能
    returnCode = DjiTest_CameraEmuMediaStartService();
    if (returnCode != DJI_ERROR_SYSTEM_MODULE_CODE_SUCCESS) {
        USER_LOG_ERROR("camera emu media init error");
}
```

2. 创建视频流处理线程

为避免因其他任务阻塞视频流处理线程，导致视频流在传输时出现花屏和绿屏的问题，需要在使用 PSDK 开发相机类负载设备时，单独创建视频流处理线程。

```
// 首先，判断网络或串口是否开启
    if (DjiPlatform_GetHalNetworkHandler() != NULL ||
DjiPlatform_GetHalUsbBulkHandler() != NULL)
```

```c
{
    // 创建名为 UserCameraMedia_SendVideoTask 的处理线程
returnCode = osalHandler->TaskCreate("user_camera_media_task",
UserCameraMedia_SendVideoTask,
2048,
                             NULL,
&s_userSendVideoThread);
if (returnCode != DJI_ERROR_SYSTEM_MODULE_CODE_SUCCESS)
{
        USER_LOG_ERROR("user send video task create error.");
        return DJI_ERROR_SYSTEM_MODULE_CODE_SUCCESS;
    }
}
```

使用 PSDK 开发的相机类负载设备在创建视频流处理线程后，需要先初始化线程状态，并向相机类负载设备申请用于缓存视频流文件的内存空间。

```c
    videoFilePath = osalHandler->Malloc(DJI_FILE_PATH_SIZE_MAX);
    if (videoFilePath == NULL) {
        USER_LOG_ERROR("malloc memory for video file path fail.");
        exit(1);
    }

    transcodedFilePath = osalHandler->Malloc(DJI_FILE_PATH_SIZE_MAX);
    if (transcodedFilePath == NULL) {
        USER_LOG_ERROR("malloc memory for transcoded file path fail.");
        exit(1);
    }

    frameInfo = osalHandler->Malloc(VIDEO_FRAME_MAX_COUNT *
sizeof(T_TestPayloadCameraVideoFrameInfo));
    if (frameInfo == NULL) {
        USER_LOG_ERROR("malloc memory for frame info fail.");
        exit(1);
    }
    memset(frameInfo, 0, VIDEO_FRAME_MAX_COUNT *
sizeof(T_TestPayloadCameraVideoFrameInfo));
```

3. 获取视频流文件信息

使用 PSDK 开发的相机类负载设备在发送视频流文件前，需读取相机类负载设备本地的 H.264 文件，以获取视频流文件信息。用户指定负载设备上准确的 H.264 文件所在的路径后，使用 PSDK 开发的相机类负载设备通过系统接口，打开用户指定的视频流文件。

```c
    returnCode = DjiUserUtil_GetCurrentFileDirPath(__FILE__,
DJI_FILE_PATH_SIZE_MAX, curFileDirPath);
    if (returnCode != DJI_ERROR_SYSTEM_MODULE_CODE_SUCCESS) {
        USER_LOG_ERROR("Get file current path error, stat = 0x%08llX", returnCode);
        exit(1);
```

```
    }
    snprintf(tempPath, DJI_FILE_PATH_SIZE_MAX, "%smedia_file/PSDK_0005.h264",
curFileDirPath);

    videoFilePath = osalHandler->Malloc(DJI_FILE_PATH_SIZE_MAX);
    if (videoFilePath == NULL) {
        USER_LOG_ERROR("malloc memory for video file path fail.");
        exit(1);
    }

    fpFile = fopen(transcodedFilePath, "rb+");
    if (fpFile == NULL) {
        USER_LOG_ERROR("open video file fail.");
        continue;
    }
```

4. 获取 H.264 文件的信息

使用 PSDK 开发的相机类负载设备通过 FFmpeg 命令读取指定的 H.264 文件的帧率、帧信息和总帧数。帧率是指相机类负载设备在 1s 内可发送的帧的数量，而帧信息是指 H.264 文件内一帧的起始位置和该帧的长度。

```
    returnCode = DjiPlayback_VideoFileTranscode(videoFilePath, "h264",
transcodedFilePath,
                                                 DJI_FILE_PATH_SIZE_MAX);
    if (returnCode != DJI_ERROR_SYSTEM_MODULE_CODE_SUCCESS) {
        USER_LOG_ERROR("transcode video file error: 0x%08llX.", returnCode);
        continue;
    }

    returnCode = DjiPlayback_GetFrameRateOfVideoFile(transcodedFilePath, &frameRate);
    if (returnCode != DJI_ERROR_SYSTEM_MODULE_CODE_SUCCESS) {
        USER_LOG_ERROR("get frame rate of video error: 0x%08llX.", returnCode);
        continue;
    }

    returnCode = DjiPlayback_GetFrameInfoOfVideoFile(transcodedFilePath,
frameInfo,  VIDEO_FRAME_MAX_COUNT, &frameCount);
    if (returnCode != DJI_ERROR_SYSTEM_MODULE_CODE_SUCCESS) {
        USER_LOG_ERROR("get frame info of video error: 0x%08llX.", returnCode);
        continue;
    }

    returnCode = DjiPlayback_GetFrameNumberByTime(frameInfo, frameCount, &frameNumber,
                                    startTimeMs);
    if (returnCode != DJI_ERROR_SYSTEM_MODULE_CODE_SUCCESS) {
        USER_LOG_ERROR("get start frame number error: 0x%08llX.", returnCode);
```

```
        continue;
    }

    if (fpFile != NULL)
        fclose(fpFile);
```

5. 视频流文件解析

使用 PSDK 开发的相机类负载设备在获取视频流文件信息后，将解析视频流文件内容，识别视频流文件的帧头。

```
dataBuffer = calloc(frameBufSize, 1);
if (dataBuffer == NULL) {
    USER_LOG_ERROR("malloc fail.");
    goto free;
}

ret = fseek(fpFile, frameInfo[frameNumber].positionInFile, SEEK_SET);
if (ret != 0) {
    USER_LOG_ERROR("fseek fail.");
    goto free;
}

dataLength = fread(dataBuffer, 1, frameInfo[frameNumber].size, fpFile);
if (dataLength != frameInfo[frameNumber].size) {
    USER_LOG_ERROR("read data from video file error.");
} else {
    USER_LOG_DEBUG("read data from video file success, len = %d B\r\n", dataLength);
}
```

6. 发送视频流文件

使用 PSDK 开发的相机类负载设备在解析视频流文件并识别视频流文件的帧头后，将以逐帧的方式发送视频流文件。使用 DJI Pilot 或基于 MSDK 开发的移动端 App 向相机类负载设备发送视频流文件传输命令后，移动端 App 将能接收并循环播放相机类负载设备中的媒体文件。

```
lengthOfDataHaveBeenSent = 0;
while (dataLength - lengthOfDataHaveBeenSent) {
    lengthOfDataToBeSent = USER_UTIL_MIN(DATA_SEND_FROM_VIDEO_STREAM_MAX_LEN,
                                dataLength - lengthOfDataHaveBeenSent);
    returnCode = DjiPayloadCamera_SendVideoStream((const uint8_t *) dataBuffer
+ lengthOfDataHaveBeenSent,
                                                    lengthOfDataToBeSent);
    if (returnCode != DJI_ERROR_SYSTEM_MODULE_CODE_SUCCESS) {
        USER_LOG_ERROR("send video stream error: 0x%08llX.", returnCode);
    }
    lengthOfDataHaveBeenSent += lengthOfDataToBeSent;
}
```

7. 调整帧率

使用 PSDK 开发的相机类负载设备能够更新视频流发送的状态，方便用户调整视频流文件传输模块传输视频的帧率。

```
returnCode = DjiPayloadCamera_GetVideoStreamState(&videoStreamState);
if (returnCode == DJI_ERROR_SYSTEM_MODULE_CODE_SUCCESS) {
    USER_LOG_DEBUG(
        "video stream state: realtimeBandwidthLimit: %d,
realtimeBandwidthBeforeFlowController: %d, realtimeBandwidthAfterFlowController:%d
busyState: %d.",
        videoStreamState.realtimeBandwidthLimit,
videoStreamState.realtimeBandwidthBeforeFlowController,
        videoStreamState.realtimeBandwidthAfterFlowController,
        videoStreamState.busyState);
} else {
    USER_LOG_ERROR("get video stream state error.");
}
```

9.5 回放下载

1. T_DjiCameraMediaDownloadPlaybackHandler 结构体

开发者根据实际的使用需求，按照 T_DjiCameraMediaDownloadPlaybackHandler 结构体构造并实现下载媒体文件原始数据、创建/下载/销毁缩略图等回放下载功能的函数，将函数注册到 PSDK 中的指定接口后，用户通过使用 DJI Pilot 或基于 MSDK 开发的移动端 App，能够从基于 PSDK 开发的相机类负载设备上下载媒体文件或实现视频流回放功能。

T_DjiCameraMediaDownloadPlaybackHandler 结构体定义如下。

```
typedef struct {
    // 实现获取媒体文件信息的功能
    T_DjiReturnCode (*GetMediaFileDir)(char *dirPath);
    T_DjiReturnCode (*GetMediaFileOriginInfo)(const char *filePath,
                         T_DjiCameraMediaFileInfo *fileInfo);
    T_DjiReturnCode (*GetMediaFileOriginData)(const char *filePath,
                                              uint32_t offset,
                            uint32_t length, uint8_t *data);

    // 实现获取媒体文件缩略图的功能
    T_DjiReturnCode (*CreateMediaFileThumbNail)(const char *filePath);
    T_DjiReturnCode (*GetMediaFileThumbNailInfo)(const char *filePath,
                               T_DjiCameraMediaFileInfo *fileInfo);
    T_DjiReturnCode (*GetMediaFileThumbNailData)(const char *filePath,
                                                  uint32_t offset,
                                                  uint32_t length,
```

```
                                                    uint8_t *data);
    T_DjiReturnCode (*DestroyMediaFileThumbNail)(const char *filePath);

    // 实现获取媒体文件截屏图的功能
    T_DjiReturnCode (*CreateMediaFileScreenNail)(const char *filePath);
    T_DjiReturnCode (*GetMediaFileScreenNailData)(const char *filePath,
                                                   uint32_t offset,
                                                   uint32_t length,
                                                   uint8_t *data);
    T_DjiReturnCode (*GetMediaFileScreenNailInfo)(const char *filePath,
                                   T_DjiCameraMediaFileInfo *fileInfo);
    T_DjiReturnCode (*DestroyMediaFileScreenNail)(const char *filePath);

    // 实现删除媒体文件的功能
    T_DjiReturnCode (*DeleteMediaFile)(char *filePath);

    // 实现控制媒体文件回放的功能
    T_DjiReturnCode (*GetMediaPlaybackStatus)(
                                   T_DjiCameraPlaybackStatus *status);
    T_DjiReturnCode (*SetMediaPlaybackFile)(const char *filePath);
    T_DjiReturnCode (*StartMediaPlayback)(void);
    T_DjiReturnCode (*StopMediaPlayback)(void);
    T_DjiReturnCode (*PauseMediaPlayback)(void);
    T_DjiReturnCode (*SeekMediaPlayback)(uint32_t playbackPosition);

    // 实现下载媒体文件的功能
    T_DjiReturnCode (*StartDownloadNotification)(void);
    T_DjiReturnCode (*StopDownloadNotification)(void);

} T_DjiCameraMediaDownloadPlaybackHandler;
```

开发回放下载功能时，使用 T_DjiCameraMediaDownloadPlaybackHandler 结构体构造并实现下载媒体文件原始数据、创建/下载/销毁缩略图等回放下载功能的函数。

使用时，首先要定义一个 T_DjiCameraMediaDownloadPlaybackHandler 结构体对象，如 s_psdkCameraMedia，接着将实现功能的函数，如 GetMediaFileDir、GetMediaFileOriginInfo、GetMediaFileOriginData、CreateMediaFileThumbNail、GetMediaFileThumbNailInfo、StartDownloadNotification 和 StopDownloadNotification 等，赋值给该结构体对象。

使用方法如下。

```
// 初始化结构体
static T_DjiCameraMediaDownloadPlaybackHandler s_psdkCameraMedia = {0};

// 实现获取媒体文件信息的功能
s_psdkCameraMedia.GetMediaFileDir = GetMediaFileDir;
s_psdkCameraMedia.GetMediaFileOriginInfo = GetMediaFileOriginInfo;
s_psdkCameraMedia.GetMediaFileOriginData = GetMediaFileOriginData;
```

```
// 实现获取媒体文件缩略图的功能
s_psdkCameraMedia.CreateMediaFileThumbNail = CreateMediaFileThumbNail;
s_psdkCameraMedia.GetMediaFileThumbNailInfo = GetMediaFileThumbNailInfo;
s_psdkCameraMedia.GetMediaFileThumbNailData = GetMediaFileThumbNailData;
s_psdkCameraMedia.DestroyMediaFileThumbNail = DestroyMediaFileThumbNail;

// 实现获取媒体文件截屏图的功能
s_psdkCameraMedia.CreateMediaFileScreenNail = CreateMediaFileScreenNail;
s_psdkCameraMedia.GetMediaFileScreenNailInfo = GetMediaFileScreenNailInfo;
s_psdkCameraMedia.GetMediaFileScreenNailData = GetMediaFileScreenNailData;
s_psdkCameraMedia.DestroyMediaFileScreenNail = DestroyMediaFileScreenNail;

// 实现删除媒体文件的功能
s_psdkCameraMedia.DeleteMediaFile = DeleteMediaFile;

// 实现控制媒体文件回放的功能
s_psdkCameraMedia.SetMediaPlaybackFile = SetMediaPlaybackFile;
s_psdkCameraMedia.StartMediaPlayback = StartMediaPlayback;
s_psdkCameraMedia.StopMediaPlayback = StopMediaPlayback;
s_psdkCameraMedia.PauseMediaPlayback = PauseMediaPlayback;
s_psdkCameraMedia.SeekMediaPlayback = SeekMediaPlayback;
s_psdkCameraMedia.GetMediaPlaybackStatus = GetMediaPlaybackStatus;

// 实现下载媒体文件的功能
s_psdkCameraMedia.StartDownloadNotification = StartDownloadNotification;
s_psdkCameraMedia.StopDownloadNotification = StopDownloadNotification;
```

最后，需要调用 DjiPayloadCamera_RegMediaDownloadPlaybackHandler 函数注册结构体对象 s_psdkCameraMedia 中定义的回放下载功能的函数。

注册方法如下。

```
returnCode =DjiPayloadCamera_RegMediaDownloadPlaybackHandler(
                                               &s_psdkCameraMedia);
if (returnCode != DJI_ERROR_SYSTEM_MODULE_CODE_SUCCESS)
{
    USER_LOG_ERROR("psdk camera media function init error.");
    return DJI_ERROR_SYSTEM_MODULE_CODE_UNKNOWN;
}
```

接下来的工作是具体实现结构体对象 s_psdkCameraMedia 中定义的回放下载功能的函数。

2. 注册回放下载功能的函数

用户可根据实际的使用需要注册回放下载功能的函数。

```
static T_DjiReturnCode StartMediaPlayback(void)
{
    T_DjiReturnCode returnCode;
```

```c
        USER_LOG_INFO("start media playback");
        returnCode = DjiPlayback_StartPlay(&s_playbackInfo);
        if (returnCode != DJI_ERROR_SYSTEM_MODULE_CODE_SUCCESS) {
            USER_LOG_ERROR("start media playback status error, stat:0x%08llX", returnCode);
            return returnCode;
        }

        return returnCode;
    }

    static T_DjiReturnCode StopMediaPlayback(void)
    {
        T_DjiReturnCode returnCode;

        USER_LOG_INFO("stop media playback");
        returnCode = DjiPlayback_StopPlay(&s_playbackInfo);
        if (returnCode != DJI_ERROR_SYSTEM_MODULE_CODE_SUCCESS) {
            USER_LOG_ERROR("stop media playback error, stat:0x%08llX", returnCode);
            return returnCode;
        }

        return returnCode;
    }

    static T_DjiReturnCode PauseMediaPlayback(void)
    {
        T_DjiReturnCode returnCode;

        USER_LOG_INFO("pause media playback");
        returnCode = DjiPlayback_PausePlay(&s_playbackInfo);
        if (returnCode != DJI_ERROR_SYSTEM_MODULE_CODE_SUCCESS) {
            USER_LOG_ERROR("pause media playback error, stat:0x%08llX", returnCode);
            return returnCode;
        }

        return returnCode;
    }

    static T_DjiReturnCode SeekMediaPlayback(uint32_t playbackPosition)
    {
        T_DjiReturnCode returnCode;

        USER_LOG_INFO("seek media playback:%d", playbackPosition);
        returnCode = DjiPlayback_SeekPlay(&s_playbackInfo, playbackPosition);
        if (returnCode != DJI_ERROR_SYSTEM_MODULE_CODE_SUCCESS) {
```

```
        USER_LOG_ERROR("seek media playback error, stat:0x%08llX", returnCode);
        return returnCode;
    }

    return returnCode;
}
```

3. 获取文件列表

1）获取媒体文件的路径

使用 PSDK 开发的相机类负载设备通过 GetMediaFileDir 接口能够获取用户指定的文件的地址，而用户使用 DJI Pilot 或基于 MSDK 开发的移动端 App 能够获取指定的媒体文件的路径。媒体文件的默认路径为 camera_emu/media_file/，用户可根据实际的使用需要，更改媒体文件所在的路径。

```
pi@raspberrypi:~/My_payload/pro_10_playback $ cd camera_emu/media_file/
pi@raspberrypi:~/My_payload/pro_10_playback/camera_emu/media_file $ tree
.
├── out.h264
├── PSDK_0001_ORG.jpg
├── PSDK_0002_ORG.jpg
├── PSDK_0003_ORG.jpg
├── PSDK_0004_ORG.mp4
└── PSDK_0005.h264
```

GetMediaFileDir 接口如下。

```
static T_DjiReturnCode GetMediaFileDir(char *dirPath)
{
    T_DjiReturnCode returnCode;
    char curFileDirPath[DJI_FILE_PATH_SIZE_MAX];
    char tempPath[DJI_FILE_PATH_SIZE_MAX];

    returnCode = DjiUserUtil_GetCurrentFileDirPath(__FILE__,
                              DJI_FILE_PATH_SIZE_MAX, curFileDirPath);
    if (returnCode != DJI_ERROR_SYSTEM_MODULE_CODE_SUCCESS) {
        USER_LOG_ERROR("Get file current path error, stat = 0x%08llX", returnCode);
        return returnCode;
    }

    snprintf(dirPath, DJI_FILE_PATH_SIZE_MAX, "%smedia_file", curFileDirPath);

    return DJI_ERROR_SYSTEM_MODULE_CODE_SUCCESS;
}
```

2）获取媒体文件的列表

指定媒体文件所在的路径后，基于 MSDK 开发的移动端 App 将向相机类负载设备发送获取媒体文件信息的命令。基于 PSDK 开发的负载设备通过调用 PSDK 中获取媒体文件信息的接口，能够获取负载设备中媒体文件的信息，如文件的名称、路径和大小等。相机类负载设备通过 FFmpeg 命令能够获取媒体文件的长度、帧率和分辨率。

```c
T_DjiReturnCode DjiTest_CameraMediaGetFileInfo(const char *filePath,
                                                T_DjiCameraMediaFileInfo *fileInfo)
{
    T_DjiReturnCode returnCode;
    T_DjiMediaFileHandle mediaFileHandle;

    returnCode = DjiMediaFile_CreateHandle(filePath, &mediaFileHandle);
    if (returnCode != DJI_ERROR_SYSTEM_MODULE_CODE_SUCCESS) {
        USER_LOG_ERROR("Media file create handle error stat:0x%08llX",
                                                            returnCode);
        return returnCode;
    }

    returnCode = DjiMediaFile_GetMediaFileType(mediaFileHandle,
                                                &fileInfo->type);
    if (returnCode != DJI_ERROR_SYSTEM_MODULE_CODE_SUCCESS) {
        USER_LOG_ERROR("Media file get type error stat:0x%08llX", returnCode);
        goto out;
    }

    returnCode = DjiMediaFile_GetMediaFileAttr(mediaFileHandle,
                                                &fileInfo->mediaFileAttr);
    if (returnCode != DJI_ERROR_SYSTEM_MODULE_CODE_SUCCESS) {
        USER_LOG_ERROR("Media file get attr error stat:0x%08llX", returnCode);
        goto out;
    }

    returnCode = DjiMediaFile_GetFileSizeOrg(mediaFileHandle,
                                                &fileInfo->fileSize);
    if (returnCode != DJI_ERROR_SYSTEM_MODULE_CODE_SUCCESS) {
        USER_LOG_ERROR("Media file get size error stat:0x%08llX", returnCode);
        goto out;
    }

out:
    returnCode = DjiMediaFile_DestroyHandle(mediaFileHandle);
    if (returnCode != DJI_ERROR_SYSTEM_MODULE_CODE_SUCCESS) {
        USER_LOG_ERROR("Media file destroy handle error stat:0x%08llX",
                                                            returnCode);
        return returnCode;
    }
```

```
    return returnCode;
}
```

3）获取媒体文件的原始数据

使用 PSDK 开发的相机类负载设备通过 GetMediaFileOriginData 接口，能够获取负载设备上的媒体文件的原始数据，而用户使用 DJI Pilot 或基于 MSDK 开发的移动端 App 能够获取指定的媒体文件所在的路径。

获取媒体文件列表的功能只能用于获取媒体文件的原始文件和缩略图。实现获取媒体文件列表的功能后，用户可以使用该功能获取指定文件的截屏图。同时，用户通过下载功能能够得到指定媒体文件的原始文件。

```c
static T_DjiReturnCode GetMediaFileOriginData(const char *filePath,
uint32_t offset,
uint32_t length,
uint8_t *data)
{
    T_DjiReturnCode returnCode;
    uint16_t realLen = 0;
    T_DjiMediaFileHandle mediaFileHandle;

    returnCode = DjiMediaFile_CreateHandle(filePath, &mediaFileHandle);
    if (returnCode != DJI_ERROR_SYSTEM_MODULE_CODE_SUCCESS) {
        USER_LOG_ERROR("Media file create handle error stat:0x%08llX",
                                                            returnCode);
        return returnCode;
    }

    returnCode = DjiMediaFile_GetDataOrg(mediaFileHandle, offset, length,
                                                        data, &realLen);
    if (returnCode != DJI_ERROR_SYSTEM_MODULE_CODE_SUCCESS) {
        USER_LOG_ERROR("Media file get data error stat:0x%08llX", returnCode);
        return returnCode;
    }

    returnCode = DjiMediaFile_DestroyHandle(mediaFileHandle);
    if (returnCode != DJI_ERROR_SYSTEM_MODULE_CODE_SUCCESS) {
        USER_LOG_ERROR("Media file destroy handle error stat:0x%08llX",
                                                            returnCode);
        return returnCode;
    }

    return DJI_ERROR_SYSTEM_MODULE_CODE_SUCCESS;
}
```

4. 创建缩略图

1）获取指定对象的路径并创建缩略图

使用 PSDK 开发的相机类负载设备通过 **CreateMediaFileThumbNail** 接口，能够获取用户指定的媒体文件的路径并创建缩略图。同时，用户使用 DJI Pilot 或基于 MSDK 开发的移动端 App 能够向负载设备发送创建指定媒体文件的缩略图的命令。

```c
static T_DjiReturnCode CreateMediaFileThumbNail(const char *filePath)
{
    T_DjiReturnCode returnCode;

    returnCode = DjiMediaFile_CreateHandle(filePath,
                                        &s_mediaFileThumbNailHandle);
    if (returnCode != DJI_ERROR_SYSTEM_MODULE_CODE_SUCCESS) {
        USER_LOG_ERROR("Media file create handle error stat:0x%08llX",
                                                        returnCode);
        return returnCode;
    }

    returnCode = DjiMediaFile_CreateThm(s_mediaFileThumbNailHandle);
    if (returnCode != DJI_ERROR_SYSTEM_MODULE_CODE_SUCCESS) {
        USER_LOG_ERROR("Media file create thumb nail error stat:0x%08llX",
                                                        returnCode);
        return returnCode;
    }

    return DJI_ERROR_SYSTEM_MODULE_CODE_SUCCESS;
}
```

2）获取缩略图的信息

使用 PSDK 开发的相机类负载设备通过 **GetMediaFileThumbNailInfo** 接口，能够获取指定媒体文件的缩略图信息。同时，用户使用 DJI Pilot 或基于 MSDK 开发的移动端 App 能够向负载设备发送获取指定媒体文件缩略图信息的命令。

```c
static T_DjiReturnCode GetMediaFileThumbNailInfo(const char *filePath,
                            T_DjiCameraMediaFileInfo *fileInfo)
{
    T_DjiReturnCode returnCode;

    USER_UTIL_UNUSED(filePath);

    if (s_mediaFileThumbNailHandle == NULL) {
        USER_LOG_ERROR("Media file thumb nail handle null error");
        return DJI_ERROR_SYSTEM_MODULE_CODE_INVALID_PARAMETER;
    }
```

```
    returnCode = DjiMediaFile_GetMediaFileType(s_mediaFileThumbNailHandle,
&fileInfo->type);
    if (returnCode != DJI_ERROR_SYSTEM_MODULE_CODE_SUCCESS) {
        USER_LOG_ERROR("Media file get type error stat:0x%08llX", returnCode);
        return returnCode;
    }

    returnCode = DjiMediaFile_GetMediaFileAttr(s_mediaFileThumbNailHandle,
&fileInfo->mediaFileAttr);
    if (returnCode != DJI_ERROR_SYSTEM_MODULE_CODE_SUCCESS) {
        USER_LOG_ERROR("Media file get attr error stat:0x%08llX", returnCode);
        return returnCode;
    }

    returnCode = DjiMediaFile_GetFileSizeThm(s_mediaFileThumbNailHandle,
&fileInfo->fileSize);
    if (returnCode != DJI_ERROR_SYSTEM_MODULE_CODE_SUCCESS) {
        USER_LOG_ERROR("Media file get size error stat:0x%08llX", returnCode);
        return returnCode;
    }

    return DJI_ERROR_SYSTEM_MODULE_CODE_SUCCESS;
}
```

3）获取缩略图

使用 PSDK 开发的相机类负载设备通过 GetMediaFileThumbNailData 接口，能够获取指定媒体文件的缩略图。同时，用户使用 DJI Pilot 或基于 MSDK 开发的移动端 App 能够向负载设备发送获取指定媒体文件缩略图的命令，并获取指定的缩略图。

```
static T_DjiReturnCode GetMediaFileThumbNailData(const char *filePath,
uint32_t offset, uint32_t length, uint8_t *data)
{
    T_DjiReturnCode returnCode;
    uint16_t realLen = 0;

    USER_UTIL_UNUSED(filePath);

    if (s_mediaFileThumbNailHandle == NULL) {
        USER_LOG_ERROR("Media file thumb nail handle null error");
        return DJI_ERROR_SYSTEM_MODULE_CODE_INVALID_PARAMETER;
    }

    returnCode = DjiMediaFile_GetDataThm(s_mediaFileThumbNailHandle, offset,
length, data, &realLen);
    if (returnCode != DJI_ERROR_SYSTEM_MODULE_CODE_SUCCESS) {
        USER_LOG_ERROR("Media file get data error stat:0x%08llX", returnCode);
        return returnCode;
```

```
    }

    return DJI_ERROR_SYSTEM_MODULE_CODE_SUCCESS;
}
```

5. 格式转换

使用 PSDK 开发的相机类负载设备通过 DjiPlayback_VideoFileTranscode 接口，能够将用户指定的媒体文件转换为 H.264 格式的文件。同时，用户使用 DJI Pilot 或基于 MSDK 开发的移动端 App 能够向负载设备发送转换媒体文件格式的指令，将指定的媒体文件转换为 H.264 格式，并获取转换格式后的媒体文件的帧率、帧信息、总帧数等信息。

```
returnCode = DjiPlayback_VideoFileTranscode(videoFilePath, "h264",
            transcodedFilePath,DJI_FILE_PATH_SIZE_MAX);
if (returnCode != DJI_ERROR_SYSTEM_MODULE_CODE_SUCCESS) {
    USER_LOG_ERROR("transcode video file error: 0x%08llX.", returnCode);
    continue;
}

returnCode = DjiPlayback_GetFrameRateOfVideoFile(transcodedFilePath,
                                                &frameRate);
if (returnCode != DJI_ERROR_SYSTEM_MODULE_CODE_SUCCESS) {
    USER_LOG_ERROR("get frame rate of video error: 0x%08llX.", returnCode);
    continue;
}

returnCode = DjiPlayback_GetFrameInfoOfVideoFile(transcodedFilePath,
                    frameInfo, VIDEO_FRAME_MAX_COUNT,&frameCount);
if (returnCode != DJI_ERROR_SYSTEM_MODULE_CODE_SUCCESS) {
    USER_LOG_ERROR("get frame info of video error: 0x%08llX.", returnCode);
    continue;
}

returnCode = DjiPlayback_GetFrameNumberByTime(frameInfo,
                    frameCount, &frameNumber,startTimeMs);
if (returnCode != DJI_ERROR_SYSTEM_MODULE_CODE_SUCCESS) {
    USER_LOG_ERROR("get start frame number error: 0x%08llX.", returnCode);
    continue;
}
```

6. 发送媒体文件

使用 PSDK 开发的相机类负载设备通过 DjiPayloadCamera_SendVideoStream 接口，能够根据转换格式后的媒体文件的帧率、帧信息、总帧数等信息找到媒体文件的帧头，并按照指定的码率向 DJI Pilot 或基于 MSDK 开发的移动端 App 发送媒体文件。同时，用户使用 DJI Pilot 或基于 MSDK 开发的移动端 App 能够接收负载设备发送的媒体文件。

```
            lengthOfDataHaveBeenSent = 0;
            while (dataLength - lengthOfDataHaveBeenSent) {
                lengthOfDataToBeSent = USER_UTIL_MIN(DATA_SEND_FROM_VIDEO_STREAM_MAX_LEN,
                        dataLength - lengthOfDataHaveBeenSent);
                returnCode = DjiPayloadCamera_SendVideoStream((const uint8_t *)
dataBuffer + lengthOfDataHaveBeenSent,lengthOfDataToBeSent);
                if (returnCode != DJI_ERROR_SYSTEM_MODULE_CODE_SUCCESS) {
                    USER_LOG_ERROR("send video stream error: 0x%08llX.", returnCode);
                }
                lengthOfDataHaveBeenSent += lengthOfDataToBeSent;
            }
```

7. 获取媒体文件的传输状态

使用 PSDK 开发的相机类负载设备通过 GetMediaPlaybackStatus 接口，能够获取当前媒体文件的传输状态。同时，用户使用 DJI Pilot 或基于 MSDK 开发的移动端 App 能够获取当前传输的媒体文件的状态信息。

```
static T_DjiReturnCode GetMediaPlaybackStatus(T_DjiCameraPlaybackStatus *status)
{
    T_DjiReturnCode returnCode;

    returnCode = DjiPlayback_GetPlaybackStatus(&s_playbackInfo, status);
    if (returnCode != DJI_ERROR_SYSTEM_MODULE_CODE_SUCCESS) {
        USER_LOG_ERROR("get playback status error, stat:0x%08llX", returnCode);
        return returnCode;
    }

    status->videoPlayProcess = (uint8_t) (((float) s_playbackInfo.playPosMs /
(float) s_playbackInfo.videoLengthMs) *100);

    USER_LOG_DEBUG("get media playback status %d %d %d %d",
                status->videoPlayProcess, status->playPosMs,
                status->videoLengthMs, status->playbackMode);

    return DJI_ERROR_SYSTEM_MODULE_CODE_SUCCESS;
}
```

9.6 代码模板

开发者在实现视频流文件传输与回放下载功能时，不需要在 main 函数中调用视频流文件传输与回放下载服务入口函数，而需要创建用户线程来实现视频流文件传输与回放下载功能。在用户线程中调用视频流文件传输与回放下载的接口，启动调度器后，视频流文件传输与回放下载接口将正常运行。

1. 代码结构

下面是实现视频流文件传输与回放下载功能的代码框架，用户只需要编写用于实现视频流文件传输与回放下载功能的 payload_cam_emu_media.c 和 payload_cam_emu_media.h 即可。

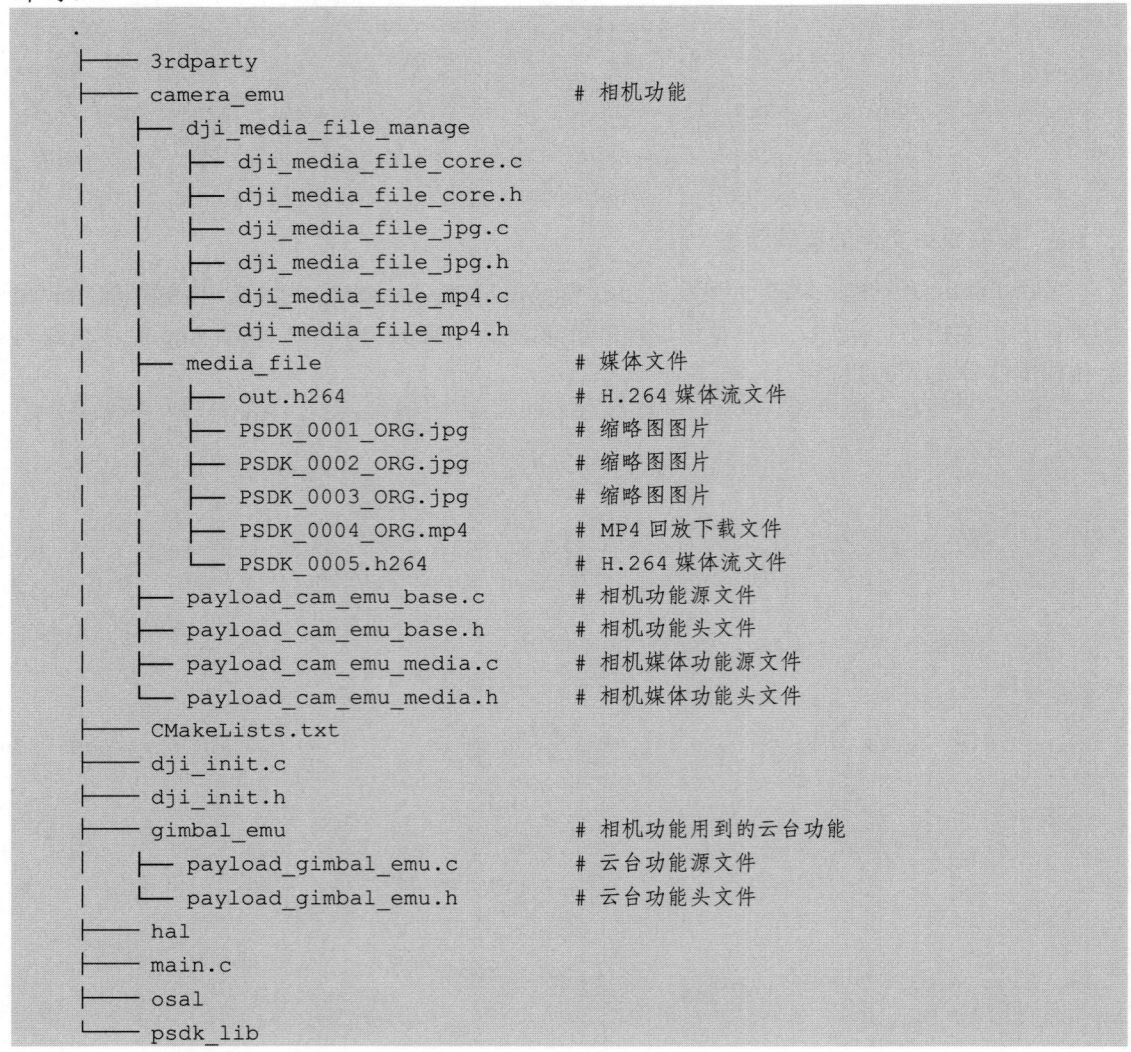

```
.
├── 3rdparty
├── camera_emu                                    # 相机功能
│   ├── dji_media_file_manage
│   │   ├── dji_media_file_core.c
│   │   ├── dji_media_file_core.h
│   │   ├── dji_media_file_jpg.c
│   │   ├── dji_media_file_jpg.h
│   │   ├── dji_media_file_mp4.c
│   │   └── dji_media_file_mp4.h
│   ├── media_file                                # 媒体文件
│   │   ├── out.h264                              # H.264 媒体流文件
│   │   ├── PSDK_0001_ORG.jpg                     # 缩略图图片
│   │   ├── PSDK_0002_ORG.jpg                     # 缩略图图片
│   │   ├── PSDK_0003_ORG.jpg                     # 缩略图图片
│   │   ├── PSDK_0004_ORG.mp4                     # MP4 回放下载文件
│   │   └── PSDK_0005.h264                        # H.264 媒体流文件
│   ├── payload_cam_emu_base.c                    # 相机功能源文件
│   ├── payload_cam_emu_base.h                    # 相机功能头文件
│   ├── payload_cam_emu_media.c                   # 相机媒体功能源文件
│   └── payload_cam_emu_media.h                   # 相机媒体功能头文件
├── CMakeLists.txt
├── dji_init.c
├── dji_init.h
├── gimbal_emu                                    # 相机功能用到的云台功能
│   ├── payload_gimbal_emu.c                      # 云台功能源文件
│   └── payload_gimbal_emu.h                      # 云台功能头文件
├── hal
├── main.c
├── osal
└── psdk_lib
```

2. 头文件模板

本任务的头文件名为 payload_cam_emu_media.h。该文件中定义的内容可满足正常的应用开发需求，如果需要实现特殊的功能，则可以按需要修改。

这里不再展示头文件代码。

3. 源文件模板

本任务的源文件名为 payload_cam_emu_media.c，以下是 payload_cam_emu_media.c 的模板文件，按照提示即可完成代码的编写。

项目9 视频流传输与回放下载

由于源文件代码行数太多，这里仅给出在 payload_cam_emu_media.c 中需要完成的部分。
（1）在服务入口函数中完成以下功能。

```
/* -- 在 DjiTest_CameraEmuMediaStartService 服务入口函数中完成以下功能 --*/
/* step 1 定义高速数据传输功能的带宽比例为 10%(数据流)、60%(视频流)和 30%(下载流)*/
/* step 2 调用 DjiHighSpeedDataChannel_SetBandwidthProportion 函数设置上述带宽比例*/
/* step 3 调用 DjiPayloadCamera_RegMediaDownloadPlaybackHandler 函数注册回放下载功能的函数*/
/* step 4 创建名为 UserCameraMedia_SendVideoTask 的视频流处理线程*/

/* ----------------------------------------------------------------*/

T_DjiReturnCode DjiTest_CameraEmuMediaStartService(void)
{
    T_DjiOsalHandler *osalHandler = DjiPlatform_GetOsalHandler();
    T_DjiReturnCode returnCode;

    USER_LOG_INFO("start DjiTest_CameraEmuMediaStartService-----------");
/* step 1 定义高速数据传输功能的带宽比例为 10%(数据流)、60%(视频流)和 30%(下载流)*/

/* ----------------------------------------------------------*

    /* step 2 调用 DjiHighSpeedDataChannel_SetBandwidthProportion 函数设置上述带宽比例*/

    /* ----------------------------------------------------*
    T_DjiAircraftInfoBaseInfo aircraftInfoBaseInfo = {0};

    if (DjiAircraftInfo_GetBaseInfo(&aircraftInfoBaseInfo) !=
DJI_ERROR_SYSTEM_MODULE_CODE_SUCCESS) {
        USER_LOG_ERROR("get aircraft information error.");
        return DJI_ERROR_SYSTEM_MODULE_CODE_UNKNOWN;
    }
    /*------按照 T_DjiCameraMediaDownloadPlaybackHandler 结构体构造并实现下载媒体文件原始数据、创建/下载/销毁缩略图等回放下载功能的函数----*/

    // 实现获取媒体文件信息的功能
    s_psdkCameraMedia.GetMediaFileDir = GetMediaFileDir;
    s_psdkCameraMedia.GetMediaFileOriginInfo = DjiTest_CameraMediaGetFileInfo;
    s_psdkCameraMedia.GetMediaFileOriginData = GetMediaFileOriginData;

    // 实现获取媒体文件缩略图的功能
    s_psdkCameraMedia.CreateMediaFileThumbNail = CreateMediaFileThumbNail;
    s_psdkCameraMedia.GetMediaFileThumbNailInfo = GetMediaFileThumbNailInfo;
    s_psdkCameraMedia.GetMediaFileThumbNailData = GetMediaFileThumbNailData;
    s_psdkCameraMedia.DestroyMediaFileThumbNail = DestroyMediaFileThumbNail;
```

```c
    // 实现获取媒体文件截屏图的功能
    s_psdkCameraMedia.CreateMediaFileScreenNail = CreateMediaFileScreenNail;
    s_psdkCameraMedia.GetMediaFileScreenNailInfo = GetMediaFileScreenNailInfo;
    s_psdkCameraMedia.GetMediaFileScreenNailData = GetMediaFileScreenNailData;
    s_psdkCameraMedia.DestroyMediaFileScreenNail = DestroyMediaFileScreenNail;

    // 实现删除媒体文件的功能
    s_psdkCameraMedia.DeleteMediaFile = DeleteMediaFile;

    // 实现控制媒体文件回放的功能
    s_psdkCameraMedia.SetMediaPlaybackFile = SetMediaPlaybackFile;
    s_psdkCameraMedia.StartMediaPlayback = StartMediaPlayback;
    s_psdkCameraMedia.StopMediaPlayback = StopMediaPlayback;
    s_psdkCameraMedia.PauseMediaPlayback = PauseMediaPlayback;
    s_psdkCameraMedia.SeekMediaPlayback = SeekMediaPlayback;
    s_psdkCameraMedia.GetMediaPlaybackStatus = GetMediaPlaybackStatus;

    // 实现下载媒体文件的功能
    s_psdkCameraMedia.StartDownloadNotification = StartDownloadNotification;
    s_psdkCameraMedia.StopDownloadNotification = StopDownloadNotification;

    if (osalHandler->MutexCreate(&s_mediaPlayCommandBufferMutex) !=
DJI_ERROR_SYSTEM_MODULE_CODE_SUCCESS) {
        USER_LOG_ERROR("mutex create error");
        return DJI_ERROR_SYSTEM_MODULE_CODE_UNKNOWN;
    }

    UtilBuffer_Init(&s_mediaPlayCommandBufferHandler,
s_mediaPlayCommandBuffer, sizeof(s_mediaPlayCommandBuffer));

    if (aircraftInfoBaseInfo.aircraftType == DJI_AIRCRAFT_TYPE_M300_RTK) {
        /* step 3 调用 DjiPayloadCamera_RegMediaDownloadPlaybackHandler 函数注册
回放下载功能的函数-----*/

    /* -----------------------------------------------------------*

    }

    if (DjiPlatform_GetHalNetworkHandler() != NULL ||
DjiPlatform_GetHalUsbBulkHandler() != NULL) {
        /* step 4 创建名为 UserCameraMedia_SendVideoTask 的视频流处理线程*/

        /* -----------------------------------------------------------*/
    }
```

```
    return DJI_ERROR_SYSTEM_MODULE_CODE_SUCCESS;
}
```

（2）在视频流处理线程中完成以下功能。

```
/* -------在视频流处理线程 UserCameraMedia_SendVideoTask 中完成以下功能-- */
/* step 1 获取视频流文件信息。读取相机类负载设备本地的 H.264 文件，获取视频流文件信息*/
/* step 2 初始化视频流处理线程，并向相机类负载设备申请用于缓存视频流文件的内存空间 */
/* step 3 获取 H.264 文件的信息。使用 FFmpeg 命令读取指定的 H.264 文件的帧率、帧信息和总帧数*/
/* step 4 视频流文件解析。获取视频流文件信息后，将解析视频流文件内容，识别视频流文件的帧头*/
/* step 5 发送视频流文件。负载设备在解析视频流文件内容并识别视频流文件的帧头后，以逐帧的方式发
送视频流文件*/

/* ---------------------------------------------------------------- */

static void *UserCameraMedia_SendVideoTask(void *arg)
{
    int ret;
    T_DjiReturnCode returnCode;
    static uint32_t sendVideoStep = 0;
    FILE *fpFile = NULL;
    unsigned long dataLength = 0;
    uint16_t lengthOfDataToBeSent = 0;
    int lengthOfDataHaveBeenSent = 0;
    char *dataBuffer = NULL;
    T_TestPayloadCameraPlaybackCommand playbackCommand = {0};
    uint16_t bufferReadSize = 0;
    char *videoFilePath = NULL;
    char *transcodedFilePath = NULL;
    float frameRate = 1.0f;
    T_TestPayloadCameraVideoFrameInfo *frameInfo = NULL;
    uint32_t frameNumber = 0;
    uint32_t frameCount = 0;
    uint32_t startTimeMs = 0;
    bool sendVideoFlag = true;
    bool sendOneTimeFlag = false;
    T_DjiDataChannelState videoStreamState = {0};
    E_DjiCameraMode mode = DJI_CAMERA_MODE_SHOOT_PHOTO;
    T_DjiOsalHandler *osalHandler = DjiPlatform_GetOsalHandler();
    uint32_t frameBufSize = 0;
    E_DjiCameraVideoStreamType videoStreamType;
    char curFileDirPath[DJI_FILE_PATH_SIZE_MAX];
    char tempPath[DJI_FILE_PATH_SIZE_MAX];

    USER_UTIL_UNUSED(arg);
    /* step 1 获取视频流文件信息。读取相机类负载设备本地的 H.264 文件，获取视频流文件信息 */

    /* ---------------------------------------------------------------- */
```

```c
    /* step 2 初始化视频流处理线程,并向相机类负载设备申请用于缓存视频流文件的内存空间 */

    /* -------------------------------------------------------------- */

    returnCode = DjiPlayback_StopPlayProcess();
    if (returnCode != DJI_ERROR_SYSTEM_MODULE_CODE_SUCCESS) {
        USER_LOG_ERROR("stop playback and start liveview error: 0x%08llX.", returnCode);
        exit(1);
    }

    while (1) {
        osalHandler->TaskSleepMs(1000 / SEND_VIDEO_TASK_FREQ);

        // response playback command
        if (osalHandler->MutexLock(s_mediaPlayCommandBufferMutex) != DJI_ERROR_SYSTEM_MODULE_CODE_SUCCESS) {
            USER_LOG_ERROR("mutex lock error");
            continue;
        }

        bufferReadSize = UtilBuffer_Get(&s_mediaPlayCommandBufferHandler, (uint8_t *) &playbackCommand,
                                       sizeof(T_TestPayloadCameraPlaybackCommand));

        if (osalHandler->MutexUnlock(s_mediaPlayCommandBufferMutex) != DJI_ERROR_SYSTEM_MODULE_CODE_SUCCESS) {
            USER_LOG_ERROR("mutex unlock error");
            continue;
        }

        if (bufferReadSize != sizeof(T_TestPayloadCameraPlaybackCommand))
            goto send;

        switch (playbackCommand.command) {
            case TEST_PAYLOAD_CAMERA_MEDIA_PLAY_COMMAND_STOP:
                snprintf(videoFilePath, DJI_FILE_PATH_SIZE_MAX, "%s", tempPath);
                startTimeMs = 0;
                sendVideoFlag = true;
                sendOneTimeFlag = false;
                break;
            case TEST_PAYLOAD_CAMERA_MEDIA_PLAY_COMMAND_PAUSE:
                sendVideoFlag = false;
                goto send;
            case TEST_PAYLOAD_CAMERA_MEDIA_PLAY_COMMAND_START:
```

```
                snprintf(videoFilePath, DJI_FILE_PATH_SIZE_MAX, "%s",
playbackCommand.path);
                startTimeMs = playbackCommand.timeMs;
                sendVideoFlag = true;
                sendOneTimeFlag = true;
                break;
            default:
                USER_LOG_ERROR("playback command invalid: %d.",
playbackCommand.command);
                sendVideoFlag = false;
                goto send;
        }
        /* step 3 获取 H.264 文件的信息。使用 FFmpeg 命令读取指定的 H.264 文件的帧率、帧
信息和总帧数*/

        /* -----------------------------------------------------------*/

        if (fpFile != NULL)
            fclose(fpFile);

        fpFile = fopen(transcodedFilePath, "rb+");

        USER_LOG_INFO("transcodedFilePath file is %s.",transcodedFilePath);

        if (fpFile == NULL) {
            USER_LOG_ERROR("open video file fail.");
            continue;
        }

send:
        if (fpFile == NULL) {
            USER_LOG_ERROR("open video file fail.");
            continue;
        }

        if (sendVideoFlag != true)
            continue;

        returnCode = DjiTest_CameraGetMode(&mode);
        if (returnCode != DJI_ERROR_SYSTEM_MODULE_CODE_SUCCESS) {
            continue;
        }

        returnCode = DjiTest_CameraGetVideoStreamType(&videoStreamType);
        if (returnCode != DJI_ERROR_SYSTEM_MODULE_CODE_SUCCESS) {
            continue;
```

```
            }

            if (mode == DJI_CAMERA_MODE_PLAYBACK && s_playbackInfo.isInPlayProcess
== false) {
                continue;
            }

            if (!USER_UTIL_IS_WORK_TURN(sendVideoStep++, frameRate,
SEND_VIDEO_TASK_FREQ))
                continue;

            frameBufSize = frameInfo[frameNumber].size;
            if (videoStreamType == DJI_CAMERA_VIDEO_STREAM_TYPE_H264_DJI_FORMAT) {
                frameBufSize = frameBufSize + VIDEO_FRAME_AUD_LEN;
            }

            /* step 4 视频流文件解析。获取视频流文件信息后，将解析视频流文件内容，识别视频流文
件的帧头*/

            /* ----------------------------------------------------------*/

            if (videoStreamType == DJI_CAMERA_VIDEO_STREAM_TYPE_H264_DJI_FORMAT) {
                memcpy(&dataBuffer[frameInfo[frameNumber].size], s_frameAudInfo,
VIDEO_FRAME_AUD_LEN);
                dataLength = dataLength + VIDEO_FRAME_AUD_LEN;
            }
            /* step 5 发送视频流文件。负载设备在解析视频流文件内容并识别视频流文件的帧头后，以
逐帧的方式发送视频流文件*/

            /* ---------------------------------------------------------- */
            if ((frameNumber++) >= frameCount) {
                USER_LOG_DEBUG("reach file tail.");
                frameNumber = 0;

                if (sendOneTimeFlag == true)
                    sendVideoFlag = false;
            }

            returnCode = DjiPayloadCamera_GetVideoStreamState(&videoStreamState);
            if (returnCode == DJI_ERROR_SYSTEM_MODULE_CODE_SUCCESS) {
                USER_LOG_DEBUG(
                    "video stream state: realtimeBandwidthLimit: %d,
realtimeBandwidthBeforeFlowController: %d, realtimeBandwidthAfterFlowController:%d
busyState: %d.",videoStreamState.realtimeBandwidthLimit,
videoStreamState.realtimeBandwidthBeforeFlowController,
    videoStreamState.realtimeBandwidthAfterFlowController,
```

```
                videoStreamState.busyState);
        } else {
            USER_LOG_ERROR("get video stream state error.");
        }

free:
        free(dataBuffer);
    }
}
```

任务设计

视频流文件传输与回放下载功能依赖高速数据传输和相机功能。本任务使用 PSDK 提供的视频流文件传输控制功能从负载设备上获取 H.264 格式的视频流文件，并在 DJI Pilot 遥控器上播放，同时使用回放下载功能从负载设备上下载 MP4 格式的视频文件，用于回放。

任务的具体设计要求如下。

（1）从 pro_9_video_playback_source_template 项目模板文件中复制一个新项目。项目的工程名为 pro_9_video_playback。

（2）本任务将实现视频流文件传输与回放下载功能，项目目录下的 camera_emu 目录用于编写代码，以实现用户线程调用视频流文件传输与回放下载功能 API 接口。

（3）以自动的方式设置网络参数。

（4）在 camera_emu 目录下编写代码，实现视频流文件传输与回放下载功能。

① 完成高速数据传输功能，设置带宽比例为 10%（数据流）、60%（视频流）和 30%（下载流）。

② 初始化相机功能和相机媒体功能。

③ 设置媒体流传输的媒体参数。

④ 设置回放下载的媒体参数。

⑤ 创建用户线程。

⑥ 在用户线程中传输 H.264 文件。

⑦ 在回放下载中控制 MP4 文件的播放、暂停和删除。

（5）使用 cmake 编译项目。

（6）执行项目。

① 测试视频流文件传输功能。

② 测试回放下载功能。

任务实施

限于篇幅，这里只给出任务实施步骤和关键代码。

1. 模板文件

连接树莓派开发板，默认路径为当前登录用户的家目录。查看当前的家目录，该目录下有一个名为 pro_9_video_playback_source_template 的目录，此目录下保存了本任务的所有模板文件。

目录结构如下。

```
pi@raspberrypi:~ $ cd pro_9_video_playback_source_template/
pi@raspberrypi:~/ pro_9_video_playback_source_template $ tree
.
├── 3rdparty
├── camera_emu                              # 相机功能
│   ├── dji_media_file_manage
│   │   ├── dji_media_file_core.c
│   │   ├── dji_media_file_core.h
│   │   ├── dji_media_file_jpg.c
│   │   ├── dji_media_file_jpg.h
│   │   ├── dji_media_file_mp4.c
│   │   └── dji_media_file_mp4.h
│   ├── media_file                          # 媒体文件
│   │   ├── out.h264                        # H.264 媒体流文件
│   │   ├── PSDK_0001_ORG.jpg               # 缩略图图片
│   │   ├── PSDK_0002_ORG.jpg               # 缩略图图片
│   │   ├── PSDK_0003_ORG.jpg               # 缩略图图片
│   │   ├── PSDK_0004_ORG.mp4               # MP4 回放下载文件
│   │   └── PSDK_0005.h264                  # H.264 媒体流文件
│   ├── payload_cam_emu_base.c              # 相机功能源文件
│   ├── payload_cam_emu_base.h              # 相机功能头文件
│   ├── payload_cam_emu_media.c             # 相机媒体功能源文件
│   └── payload_cam_emu_media.h             # 相机媒体功能头文件
├── CMakeLists.txt
├── dji_init.c
├── dji_init.h
├── gimbal_emu                              # 相机功能用到的云台功能
│   ├── payload_gimbal_emu.c                # 云台功能源文件
│   └── payload_gimbal_emu.h                # 云台功能头文件
├── hal
├── main.c
├── osal
└── psdk_lib
```

2. 复制模板目录

将 pro_9_video_playback_source_template 目录下的所有文件和目录复制到 pro_9_video_playback 目录中。

```
pi@raspberrypi:~ $ cp -r pro_9_video_playback_source_template pro_9_video_playback
```

3. 修改 dji_init.c

（1）填写用户账号信息。

```
#define USER_App_NAME                   "RaspberryPi"
#define USER_App_ID                     "126413"
#define USER_App_KEY                    "0ea2eadb314d1a6e870975364e2bf4a"
#define USER_App_LICENSE                "HZ/GptycWQ………………"
#define USER_DEVELOPER_ACCOUNT          "abc@outlo**.com"
#define USER_BAUD_RATE                  "460800"
```

（2）开启网络。

```
/*--在通信参数中开启串口和网络功能，定义 CONFIG_HARDWARE_CONNECTION 为
DJI_USE_UART_AND_NETWORK_DEVICE-----------*/
/*---------定义通信参数---------*/
#define DJI_USE_ONLY_UART                       (0)
#define DJI_USE_UART_AND_USB_BULK_DEVICE        (1)
#define DJI_USE_UART_AND_NETWORK_DEVICE         (2)

#define CONFIG_HARDWARE_CONNECTION      DJI_USE_UART_AND_NETWORK_DEVICE
/* ------------------------------------------------------------------*/
```

4. 修改 CMakeLists.txt

在 CMakeLists.txt 中需要修改 3 处，分别是修改项目的工程名、添加相机功能的目录和文件。按照下文提示找到相应的位置进行修改即可。

（1）设置本项目的工程名为 pro_9_video_playback。

```
# --------------------------设置工程名------------------------------
project(pro_9_video_playback C)
# ----------------------------------------------------------------
```

（2）添加相机功能的目录和文件。

```
# ------在此处添加相机功能源文件 camera_emu/*.c 和目录 ./camera_emut--------
file(GLOB_RECURSE MODULE_SAMPLE_SRC camera_emu/*.c)
include_directories(./camera_emu)
# ----------------------------------------------------------------
```

5. 修改主函数 main.c

（1）添加头文件。

```
/* --添加相机功能头文件和媒体功能头文件-------------------------*/
#include <camera_emu/payload_cam_emu_media.h>
#include <camera_emu/payload_cam_emu_base.h>
/* ------------------------------------------------------*/
```

（2）添加相机功能和媒体功能服务入口函数。

```
int main(int argc, char **argv)
{

    /* ----------在此处完成初始化相机功能和相机媒体功能-----------------*/
```

```
            /* step 1 调用 DjiTest_CameraEmuBaseStartService 函数初始化相机功能---*/
            returnCode = DjiTest_CameraEmuBaseStartService();
            if (returnCode != DJI_ERROR_SYSTEM_MODULE_CODE_SUCCESS) {
                USER_LOG_ERROR("camera emu common init error");
            }
            /* ----------------------------------------------------------------*/

            /* step 2 调用 DjiTest_CameraEmuMediaStartService 函数初始化相机媒体功能---*/
            returnCode = DjiTest_CameraEmuMediaStartService();
            if (returnCode != DJI_ERROR_SYSTEM_MODULE_CODE_SUCCESS) {
                USER_LOG_ERROR("camera emu media init error");
            }
            /* ----------------------------------------------------------------*/
    }
```

6. 修改 payload_cam_emu_media.h

相机媒体功能头文件 payload_cam_emu_media.h 无须修改。

7. 修改 payload_cam_emu_media.c

根据 payload_cam_emu_media.c 模板文件，需要完成相应的功能代码。

（1）服务入口函数。

在服务入口函数 DjiTest_CameraEmuMediaStartService 中完成以下功能代码。

```
    T_DjiReturnCode DjiTest_CameraEmuMediaStartService(void)
    {

/* step 1 定义高速数据传输功能的带宽比例为10%(数据流)、60%(视频流)和30%(下载流)*/
        const T_DjiDataChannelBandwidthProportionOfHighspeedChannel
bandwidthProportionOfHighspeedChannel = {10, 60, 30};
    /* ----------------------------------------------------------------*/

/* step 2 调用 DjiHighSpeedDataChannel_SetBandwidthProportion 函数设置上述带宽比例*/
        returnCode = DjiHighSpeedDataChannel_SetBandwidthProportion
(bandwidthProportionOfHighspeedChannel);
        if (returnCode != DJI_ERROR_SYSTEM_MODULE_CODE_SUCCESS) {
            USER_LOG_ERROR("Set data channel bandwidth width proportion error.");
            return DJI_ERROR_SYSTEM_MODULE_CODE_UNKNOWN;
        }
    /* ----------------------------------------------------------------*/

    /*-----按照 T_DjiCameraMediaDownloadPlaybackHandler 结构体构造实现并下载媒体文件原始数
据、创建/下载/销毁缩略图等回放下载功能的函数-*/

            if (aircraftInfoBaseInfo.aircraftType == DJI_AIRCRAFT_TYPE_M300_RTK) {
/* step 3 调用 DjiPayloadCamera_RegMediaDownloadPlaybackHandler 函数注册回放下载功能的函数*/
```

```
                returnCode = DjiPayloadCamera_RegMediaDownloadPlaybackHandler
                                                            (&s_psdkCameraMedia);
            if (returnCode != DJI_ERROR_SYSTEM_MODULE_CODE_SUCCESS) {
                USER_LOG_ERROR("psdk camera media function init error.");
                return DJI_ERROR_SYSTEM_MODULE_CODE_UNKNOWN;
            }
/* ----------------------------------------------------------------*/
        }
    /* step 4 创建名为UserCameraMedia_SendVideoTask的视频流处理线程*/
            returnCode = osalHandler->TaskCreate("user_camera_media_task",
UserCameraMedia_SendVideoTask, 2048,NULL, &s_userSendVideoThread);
            if (returnCode != DJI_ERROR_SYSTEM_MODULE_CODE_SUCCESS) {
                USER_LOG_ERROR("user send video task create error.");
                return DJI_ERROR_SYSTEM_MODULE_CODE_SUCCESS;
            }
    }
    /* ----------------------------------------------------------------*/
```

（2）视频流处理线程。

在视频流处理线程 UserCameraMedia_SendVideoTask 中完成以下功能代码。

```
    static void *UserCameraMedia_SendVideoTask(void *arg)
    {
     /* step 1 获取视频流文件信息。读取相机类负载设备本地的H.264文件，获取视频流文件信息 */
        returnCode = DjiUserUtil_GetCurrentFileDirPath(__FILE__,
DJI_FILE_PATH_SIZE_MAX, curFileDirPath);
        if (returnCode != DJI_ERROR_SYSTEM_MODULE_CODE_SUCCESS) {
            USER_LOG_ERROR("Get file current path error, stat = 0x%08llX", returnCode);
            exit(1);
        }
        if (s_isMediaFileDirPathConfigured == true) {
            snprintf(tempPath, DJI_FILE_PATH_SIZE_MAX, "%sPSDK_0005.h264",
s_mediaFileDirPath);
        } else {
            snprintf(tempPath, DJI_FILE_PATH_SIZE_MAX,
"%smedia_file/PSDK_0005.h264", curFileDirPath);
        }
        /* ---------------------------------------------------------------- */

     /* step 2 初始化视频流处理线程，并向相机类负载设备申请用于缓存视频流文件的内存空间*/
        videoFilePath = osalHandler->Malloc(DJI_FILE_PATH_SIZE_MAX);
        if (videoFilePath == NULL) {
            USER_LOG_ERROR("malloc memory for video file path fail.");
            exit(1);
```

```c
        }

        transcodedFilePath = osalHandler->Malloc(DJI_FILE_PATH_SIZE_MAX);
        if (transcodedFilePath == NULL) {
            USER_LOG_ERROR("malloc memory for transcoded file path fail.");
            exit(1);
        }

        frameInfo = osalHandler->Malloc(VIDEO_FRAME_MAX_COUNT *
sizeof(T_TestPayloadCameraVideoFrameInfo));
        if (frameInfo == NULL) {
            USER_LOG_ERROR("malloc memory for frame info fail.");
            exit(1);
        }
        memset(frameInfo, 0, VIDEO_FRAME_MAX_COUNT *
sizeof(T_TestPayloadCameraVideoFrameInfo));
        /* -------------------------------------------------------------- */

/* step 3 获取H.264文件的信息。使用FFmpeg命令读取指定的H.264文件的帧率、帧信息和总帧数*/
        returnCode = DjiPlayback_VideoFileTranscode(videoFilePath, "h264",
transcodedFilePath,DJI_FILE_PATH_SIZE_MAX);
        if (returnCode != DJI_ERROR_SYSTEM_MODULE_CODE_SUCCESS) {
            USER_LOG_ERROR("transcode video file error: 0x%08llX.", returnCode);
            continue;
        }

        returnCode = DjiPlayback_GetFrameRateOfVideoFile(transcodedFilePath,
&frameRate);
        if (returnCode != DJI_ERROR_SYSTEM_MODULE_CODE_SUCCESS) {
            USER_LOG_ERROR("get frame rate of video error: 0x%08llX.", returnCode);
            continue;
        }
        returnCode = DjiPlayback_GetFrameInfoOfVideoFile(transcodedFilePath,
frameInfo, VIDEO_FRAME_MAX_COUNT,&frameCount);
        if (returnCode != DJI_ERROR_SYSTEM_MODULE_CODE_SUCCESS) {
            USER_LOG_ERROR("get frame info of video error: 0x%08llX.", returnCode);
            continue;
        }
        returnCode = DjiPlayback_GetFrameNumberByTime(frameInfo, frameCount,
&frameNumber,startTimeMs);
        if (returnCode != DJI_ERROR_SYSTEM_MODULE_CODE_SUCCESS) {
            USER_LOG_ERROR("get start frame number error: 0x%08llX.", returnCode);
            continue;
        }
        /* --------------------------------------------------------------*/
```

```
/* step 4 视频流文件解析。获取视频流文件信息后,将解析视频流文件内容,识别视频流文件的帧头*/
            dataBuffer = calloc(frameBufSize, 1);
            if (dataBuffer == NULL) {
                USER_LOG_ERROR("malloc fail.");
                goto free;
            }

            ret = fseek(fpFile, frameInfo[frameNumber].positionInFile, SEEK_SET);
            if (ret != 0) {
                USER_LOG_ERROR("fseek fail.");
                goto free;
            }

            dataLength = fread(dataBuffer, 1, frameInfo[frameNumber].size, fpFile);
            if (dataLength != frameInfo[frameNumber].size) {
                USER_LOG_ERROR("read data from video file error.");
            } else {
                USER_LOG_DEBUG("read data from video file success, len = %d B\r\n", dataLength);
            }
            /* ---------------------------------------------------------*/

/* step 5 发送视频流文件。负载设备在解析视频流文件内容并识别视频流文件的帧头后,以逐帧的方式发送视频流文件*/
            lengthOfDataHaveBeenSent = 0;
            while (dataLength - lengthOfDataHaveBeenSent) {
                lengthOfDataToBeSent = USER_UTIL_MIN(DATA_SEND_FROM_VIDEO_STREAM_MAX_LEN,
                                                    dataLength - lengthOfDataHaveBeenSent);
                returnCode = DjiPayloadCamera_SendVideoStream((const uint8_t *) dataBuffer + lengthOfDataHaveBeenSent,
                                                              lengthOfDataToBeSent);
                if (returnCode != DJI_ERROR_SYSTEM_MODULE_CODE_SUCCESS) {
                    USER_LOG_ERROR("send video stream error: 0x%08llX.", returnCode);
                }
                lengthOfDataHaveBeenSent += lengthOfDataToBeSent;
            }
            /* ---------------------------------------------------------- */
```

8. 编译并执行项目

```
# 在项目pro_9_video_playback目录下,创建build目录,并进入build目录
pi@raspberrypi:~/ pro_9_video_playback$ mkdir build && cd build
```

```
# 使用 cmake 创建项目编译环境
pi@raspberrypi:~/ pro_9_video_playback/build $ cmake ..

# make 编译
pi@raspberrypi:~/ pro_9_video_playback/build $ make

# 执行文件位于 build/bin 目录下
pi@raspberrypi:~/ pro_9_video_playback/build $cd bin/

# 使用 sudo 执行项目的执行文件
pi@raspberrypi:~/ pro_9_video_playback/build/bin $ sudo ./ pro_9_video_playback
```

任务评价

任务过程评价表

任务实施人姓名_____ 学号_____ 时间_____

评价项目及标准		分值	小组评议	教师评议
技术能力	1. 对基本概念的熟悉程度	10		
	2. 掌握启用网络功能的方法	10		
	3. 掌握 PSDK 的高速数据传输功能	10		
	4. 掌握 PSDK 的视频流文件传输功能	10		
	5. 掌握 PSDK 的回放下载功能	10		
	6. 掌握视频流文件传输与回放下载功能的测试方法	10		
执行能力	1. 出勤情况	5		
	2. 遵守纪律情况	5		
	3. 是否主动参与，有无提问记录	5		
	4. 有无职业意识	5		
社会能力	1. 能否有效沟通	5		
	2. 是否使用基本的文明礼貌用语	5		
	3. 能否与组员主动交流、积极合作	5		
	4. 能否自我学习、自我管理	5		
总分		100		
评定等级：				
评价意见			学习意见	

评定等级具体如下。

A：优，得分>90。B：好，90≥得分>80。C：一般，80≥得分>60。D：有待提高，得分≤60。

小结

本任务首先简要介绍了 DJI 无人机数据传输、配置网络参数、视频流文件传输及回放下载的相关知识；其次介绍了与视频流文件传输和回放下载相关的处理视频流、下载流的基本功能；最后介绍了视频流文件传输与回放下载功能的代码框架。

任务 2　实时视频流传输与回放下载

➡ 任务要求

本任务使用 PSDK 提供的视频流传输控制功能从负载设备（树莓派）上获取 H.264 格式的实时视频流，并在 DJI Pilot 遥控器上播放。

➡ 知识导入

9.7　概述

本任务涉及 PSDK 的多个功能，包括高速数据传输、视频流传输、回放下载、树莓派摄像头、实时 H.264 数据流，以及相机拍照功能和云台功能等。

关于高速数据传输、视频流传输、回放下载的相关知识请参阅"项目 9 任务 1 视频流文件传输与回放下载"中的"知识导入"部分。

9.8　H.264 编码

H.264 是一种高性能的视频编解码技术。所谓视频编码方式是指通过压缩技术，将原始视频格式文件转换成另一种视频格式文件的方式。目前，国际上制定视频编解码技术的组织有两个，一个是"国际电联（ITU-T）"，它制定的标准有 H.261、H.263、H.263+等，这些标准被统称为 H.26X 系列，主要应用于实时视频通信领域，如会议电视、可视电话等；另一个是"国际标准化组织（ISO）"，它制定的标准有 MPEG-1、MPEG-2、MPEG-4、MPEG-7、MPEG-21，这些标准被统称为 MPEG 系列。

H.264 是由两个组织联合组建的联合视频组（JVT）制定的新数字视频编码标准，所以它既是 ITU-T 的 H.264，又是 ISO/IEC 的 MPEG-4 高级视频编码，而且它将成为 MPEG-4 标准的第 10 部分。因此，不论是 MPEG-4 AVC、MPEG-4 Part 10，还是 ISO/IEC 14496-10，都是指 H.264。

视频是连续的图像序列，由连续的帧构成，一帧即为一幅图像。由于人眼的视觉暂留效应，当帧序列以一定的速率播放时，我们看到的就是动作连续的视频。

H.264 编码框架分为两层。

（1）VCL（Video Coding Layer，视频编码层）：负责高效的视频内容表示，将输入的流进行编码。VCL 是管理 H.264 的视频数据层，以实现更高的视频压缩比。

（2）NAL（Network Abstraction Layer，网络提取层）：负责将编码完成的内容打包成符合 H.264 格式的数据流，以网络所要求的恰当的方式对数据流进行打包和传送。H.264 编码框架如图 9-5 所示。

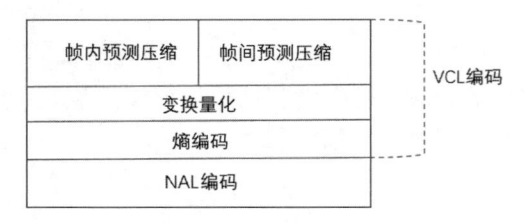

图 9-5　H.264 编码框架

在不考虑 H.264 压缩算法细节的情况下，VCL 的主要功能是压缩、切分数据，压缩、切分后的 VCL 数据会被包装成 NAL 中的一部分。

NAL 与 H.264 压缩算法无关，NAL 可实现良好的网络亲和性，即适用于各种传输网络。NAL 由 header 和 payload 两个部分组成。

其中，header 部分用于存储标志信息，如 NALU 的类型。NAL 会打包 VCL 数据，但是这并不意味着所有的 NALU 负载的都是 VCL，也有一些 NALU 仅仅存储了与编解码信息相关的数据；而 payload 部分用于存储真正的数据。

本任务使用的 H.264 码流处理部分已经被封装成树莓派的库文件，直接调用即可。

```
├── raspi_camera                    # 实现树莓派相机功能的目录
│   ├── include
│   │   └── RaspiCam.h              # 树莓派相机库的头文件
│   └── lib
│       └── libRaspiCamera.a        # 树莓派相机库的库文件
```

头文件 RaspiCam.h 中定义了视频参数结构体 RASPI_VID_PARAMS 和设置视频参数的函数 main_process，以及设置回调函数 raspi_vid_callback_set 的函数，回调函数在 payload_cam_emu_media.c 文件中定义，名为 PsdkTest_RaspiCamH264Callback，其功能是自动接收每个树莓派相机的 NALU 单元数据，推送当前帧数据。

```
#ifndef RASPI_CAM_H
#define RASPI_CAM_H

#include <stdint.h>

#ifdef __cplusplus
extern "C" {
#endif
```

```
typedef struct {
  int width;            // 图像宽
  int height;           // 图像高
  int framerate;        // 帧率，即每秒视频中有多少帧图像
  int addAud;           // 1 表示 enable, 0 表示 disable
}RASPI_VID_PARAMS;

// 用于设置回调函数
void raspi_vid_callback_set(void(*cb)(const char* h264_data, uint32_t data_len));

// 用于设置图像参数
int main_process(RASPI_VID_PARAMS *params);

#ifdef __cplusplus
}
#endif

#endif // RASPI_CAM_H
```

9.9 代码模板

当开发者实现实时视频流传输与回放下载功能时，不需要在 main 函数中调用视频流文件传输与回放下载服务入口函数，而需要创建用户线程来实现实时视频流传输与回放下载功能。在用户线程中调用实时视频流传输与回放下载功能的接口，启动调度器后，实时视频流传输与回放下载接口将正常运行。

1. 代码结构

下面是实现实时视频流传输与回放下载功能的代码框架，用户只需要编写用于实现实时视频流传输与回放下载功能的 payload_cam_emu_media.c 和 payload_cam_emu_media.h 即可。

```
.
├── 3rdparty
├── camera_emu                          # 相机功能
│   ├── dji_media_file_manage
│   │   ├── dji_media_file_core.c
│   │   ├── dji_media_file_core.h
│   │   ├── dji_media_file_jpg.c
│   │   ├── dji_media_file_jpg.h
│   │   ├── dji_media_file_mp4.c
│   │   └── dji_media_file_mp4.h
│   ├── media_file                      # 媒体文件
│   │   ├── out.h264                    # H.264 媒体流文件
```

```
        |     |     ├── PSDK_0001_ORG.jpg       # 缩略图图片
        |     |     ├── PSDK_0002_ORG.jpg       # 缩略图图片
        |     |     ├── PSDK_0003_ORG.jpg       # 缩略图图片
        |     |     ├── PSDK_0004_ORG.mp4       # MP4 回放下载文件
        |     |     └── PSDK_0005.h264          # H.264 媒体流文件
        |     ├── payload_cam_emu_base.c        # 相机功能源文件
        |     ├── payload_cam_emu_base.h        # 相机功能头文件
        |     ├── payload_cam_emu_media.c       # 相机媒体功能源文件
        |     └── payload_cam_emu_media.h       # 相机媒体功能头文件
        ├── CMakeLists.txt
        ├── dji_init.c
        ├── dji_init.h
        ├── gimbal_emu                          # 相机功能用到的云台功能
        |     ├── payload_gimbal_emu.c          # 云台功能源文件
        |     └── payload_gimbal_emu.h          # 云台功能头文件
        ├── raspi_camera                        # 实现树莓派相机功能的目录
        |     ├── include
        |     |     └── RaspiCam.h              # 树莓派相机库的头文件
        |     └── lib
        |           └── libRaspiCamera.a        # 树莓派相机库的库文件
        ├── hal
        ├── main.c
        ├── osal
        └── psdk_lib
```

2. 头文件模板

本任务的头文件名为 payload_cam_emu_media.h。该文件定义的内容可满足正常的应用开发需求，如果需要实现特殊的功能，则可以按需要修改。

这里不再展示头文件代码。

3. 源文件模板

本任务的源文件名为 payload_cam_emu_media.c，以下是 payload_cam_emu_media.c 的模板文件，按照提示即可完成代码的编写。

由于源文件代码行数太多，这里仅给出在 payload_cam_emu_media.c 中需要完成的功能。

在服务入口函数中完成以下功能。

```
/* -- 在 DjiTest_CameraEmuMediaStartService 服务入口函数中完成以下功能--*/
/* step 1 定义高速数据传输功能的带宽比例为 10%(数据流)、60%(视频流)和 30%(下载流)*/
/* step 2 调用 DjiHighSpeedDataChannel_SetBandwidthProportion 函数设置上述带宽比例*/
/* step 3 调用 DjiPayloadCamera_RegMediaDownloadPlaybackHandler 函数注册回放下载功能的函数*/
/* step 4 创建名为 UserCameraMedia_SendRaspiCamVideoTask 的视频流处理线程*/

/* ---------------------------------------------------------------- */

    T_DjiReturnCode DjiTest_CameraEmuMediaStartService(void)
    {
```

```c
    T_DjiOsalHandler *osalHandler = DjiPlatform_GetOsalHandler();
    T_DjiReturnCode returnCode;

    /* step 1 定义高速数据传输功能的带宽比例为10%(数据流)、60%(视频流)和30%(下载流)*/

    /* -----------------------------------------------------------*/
/* step 2 调用DjiHighSpeedDataChannel_SetBandwidthProportion函数设置上述带宽比例*/

    /* -----------------------------------------------------------*/

    T_DjiAircraftInfoBaseInfo aircraftInfoBaseInfo = {0};

    if (DjiAircraftInfo_GetBaseInfo(&aircraftInfoBaseInfo) !=
DJI_ERROR_SYSTEM_MODULE_CODE_SUCCESS) {
        USER_LOG_ERROR("get aircraft information error.");
        return DJI_ERROR_SYSTEM_MODULE_CODE_UNKNOWN;
    }

    s_psdkCameraMedia.GetMediaFileDir = GetMediaFileDir;
    s_psdkCameraMedia.GetMediaFileOriginInfo = DjiTest_CameraMediaGetFileInfo;
    s_psdkCameraMedia.GetMediaFileOriginData = GetMediaFileOriginData;

    s_psdkCameraMedia.CreateMediaFileThumbNail = CreateMediaFileThumbNail;
    s_psdkCameraMedia.GetMediaFileThumbNailInfo = GetMediaFileThumbNailInfo;
    s_psdkCameraMedia.GetMediaFileThumbNailData = GetMediaFileThumbNailData;
    s_psdkCameraMedia.DestroyMediaFileThumbNail = DestroyMediaFileThumbNail;

    s_psdkCameraMedia.CreateMediaFileScreenNail = CreateMediaFileScreenNail;
    s_psdkCameraMedia.GetMediaFileScreenNailInfo = GetMediaFileScreenNailInfo;
    s_psdkCameraMedia.GetMediaFileScreenNailData = GetMediaFileScreenNailData;
    s_psdkCameraMedia.DestroyMediaFileScreenNail = DestroyMediaFileScreenNail;

    s_psdkCameraMedia.DeleteMediaFile = DeleteMediaFile;
    s_psdkCameraMedia.SetMediaPlaybackFile = SetMediaPlaybackFile;
    s_psdkCameraMedia.StartMediaPlayback = StartMediaPlayback;
    s_psdkCameraMedia.StopMediaPlayback = StopMediaPlayback;
    s_psdkCameraMedia.PauseMediaPlayback = PauseMediaPlayback;
    s_psdkCameraMedia.SeekMediaPlayback = SeekMediaPlayback;
    s_psdkCameraMedia.GetMediaPlaybackStatus = GetMediaPlaybackStatus;

    s_psdkCameraMedia.StartDownloadNotification = StartDownloadNotification;
    s_psdkCameraMedia.StopDownloadNotification = StopDownloadNotification;

    if (osalHandler->MutexCreate(&s_mediaPlayCommandBufferMutex) !=
DJI_ERROR_SYSTEM_MODULE_CODE_SUCCESS) {
```

```
            USER_LOG_ERROR("mutex create error");
            return DJI_ERROR_SYSTEM_MODULE_CODE_UNKNOWN;
    }

    UtilBuffer_Init(&s_mediaPlayCommandBufferHandler,
s_mediaPlayCommandBuffer, sizeof(s_mediaPlayCommandBuffer));

    if (aircraftInfoBaseInfo.aircraftType == DJI_AIRCRAFT_TYPE_M300_RTK) {
/* step 3 调用 DjiPayloadCamera_RegMediaDownloadPlaybackHandler 函数注册回放下载功能的
函数-----*/

/* ------------------------------------------------------------*/
    }
    if (DjiPlatform_GetHalNetworkHandler() != NULL ||
DjiPlatform_GetHalUsbBulkHandler() != NULL) {
    /* step 4 创建名为 UserCameraMedia_SendRaspiCamVideoTask 的视频流处理线程----*/


/* ------------------------------------------------------------*/

    }

    return DJI_ERROR_SYSTEM_MODULE_CODE_SUCCESS;
}

/* --在视频流处理线程 UserCameraMedia_SendRaspiCamVideoTask 中完成以下功能------- */
/* step 1 设置视频参数，分辨率为 1280 × 720，每秒 30 帧*/
/* step 2 在启动实时视频流传输前，先停止可能存在的回放*/
/* step 3 设置 H.264 视频流接收函数 (PsdkTest_RaspiCamH264Callback 回调函数)*/
/* step 4 调用树莓派 main_process 函数控制摄像头*/

/* ---------------------------------------------------------------- */

static void *UserCameraMedia_SendRaspiCamVideoTask(void *arg)
{
    int ret;
    T_DjiReturnCode psdkStat;
    RASPI_VID_PARAMS camera_param;

    /* step 1 设置视频参数，分辨率为 1280 × 720，每秒 30 帧*/

    camera_param.addAud = 1;
    /* ----------------------------------------------*/

    /* step 2 在启动实时视频流传输前，先停止可能存在的回放*/
```

```
/* ------------------------------------------------*/

/* step 3 设置 H.264 视频流接收函数(PsdkTest_RaspiCamH264Callback 回调函数)*/


/* ------------------------------------------------*/

/* step 4 调用树莓派 main_process 函数控制摄像头*/


/* ------------------------------------------------*/

// 异常退出
USER_LOG_ERROR(" Raspi cam run out by some error.");
}
```

任务设计

实时视频流传输与回放下载功能依赖高速数据传输和相机功能。本任务使用 PSDK 提供的实时视频流传输控制功能从负载设备（树莓派+摄像头）上获取 H.264 格式的实时视频流，并在 DJI Pilot 遥控器上播放，同时使用回放下载功能从负载设备上下载 MP4 格式的视频文件，用于回放。（注意：本任务的重点是实时视频传输）

任务的具体设计要求如下。

（1）从 pro_9_realtime_video__source_template 项目模板文件中复制一个新项目。项目的工程名为 pro_9_ realtime_video。

（2）本任务将实现实时视频流传输与回放下载功能，项目目录下的 camera_emu 目录用于编写代码，以实现用户线程调用实时视频流传输与回放下载功能 API 接口。

（3）以自动的方式设置网络参数。

（4）在 camera_emu 目录下编写代码，实现实时视频流传输与回放下载功能。

① 完成高速数据传输功能，设置带宽比例为 10%（数据流）、60%（视频流）和 30%（下载流）。

② 初始化相机功能和相机媒体功能。

③ 设置媒体流传输的媒体参数。

④ 设置回放下载的媒体参数。

⑤ 创建用户线程。

⑥ 在用户线程中传输 H.264 实时视频流。

⑦ 在回放下载中控制 MP4 文件的播放、暂停和删除。

（5）使用 cmake 编译项目。

（6）执行项目。

① 测试实时视频流传输功能。

② 测试回放下载功能。

➡ 任务实施

限于篇幅，这里只给出任务实施步骤和关键代码。

1. 模板文件

连接树莓派开发板，默认路径为当前登录用户的家目录。查看当前的家目录，该目录下有一个名为 pro_9_realtime_video _source_template 的目录，此目录下保存了本任务的所有模板文件。

目录结构如下。

```
pi@raspberrypi:~ $ cd pro_9_realtime_video_source_template/
pi@raspberrypi:~/ pro_9_realtime_video_source_template/ $ tree
.
├── 3rdparty
├── camera_emu                              # 相机功能
│   ├── dji_media_file_manage
│   │   ├── dji_media_file_core.c
│   │   ├── dji_media_file_core.h
│   │   ├── dji_media_file_jpg.c
│   │   ├── dji_media_file_jpg.h
│   │   ├── dji_media_file_mp4.c
│   │   └── dji_media_file_mp4.h
│   ├── media_file                          # 媒体文件
│   │   ├── out.h264                        # H.264 媒体流文件
│   │   ├── PSDK_0001_ORG.jpg               # 缩略图图片
│   │   ├── PSDK_0002_ORG.jpg               # 缩略图图片
│   │   ├── PSDK_0003_ORG.jpg               # 缩略图图片
│   │   ├── PSDK_0004_ORG.mp4               # MP4 回放下载文件
│   │   └── PSDK_0005.h264                  # H.264 媒体流文件
│   ├── payload_cam_emu_base.c              # 相机功能源文件
│   ├── payload_cam_emu_base.h              # 相机功能头文件
│   ├── payload_cam_emu_media.c             # 相机媒体功能源文件
│   └── payload_cam_emu_media.h             # 相机媒体功能头文件
├── CMakeLists.txt
├── dji_init.c
├── dji_init.h
├── gimbal_emu                              # 相机功能用到的云台功能
│   ├── payload_gimbal_emu.c                # 云台功能源文件
│   └── payload_gimbal_emu.h                # 云台功能头文件
├── raspi_camera                            # 实现树莓派相机功能的目录
│   ├── include
│   │   └── RaspiCam.h                      # 树莓派相机库的头文件
│   └── lib
│       └── libRaspiCamera.a                # 树莓派相机库的库文件
├── hal
├── main.c
```

```
├── osal
└── psdk_lib
```

2. 复制模板目录

将 pro_9_realtime_video_source_template 目录下的所有文件和目录复制到 pro_9_realtime_video 目录中。

```
pi@raspberrypi:~ $ cp -r pro_9_realtime_video _source_template pro_9_realtime_video
```

3. 修改 dji_init.c

（1）填写用户账号信息。

```
#define USER_App_NAME              "RaspberryPi"
#define USER_App_ID                "126413"
#define USER_App_KEY               "0ea2eadb314d1a6e870975364e2bf4a"
#define USER_App_LICENSE           "HZ/GptycWQ........................"
#define USER_DEVELOPER_ACCOUNT     "abc@outlo**.com"
#define USER_BAUD_RATE             "460800"
```

（2）开启网络。

```
/*--在通信参数中开启串口和网络功能,定义 CONFIG_HARDWARE_CONNECTION 为
DJI_USE_UART_AND_NETWORK_DEVICE*/
/*---------定义通信参数--------*/
#define DJI_USE_ONLY_UART                    (0)
#define DJI_USE_UART_AND_USB_BULK_DEVICE     (1)
#define DJI_USE_UART_AND_NETWORK_DEVICE      (2)

#define CONFIG_HARDWARE_CONNECTION     DJI_USE_UART_AND_NETWORK_DEVICE
/* -----------------------------------------------------------------*/
```

4. 修改 CMakeLists.txt

在 CMakeLists.txt 中需要设置项目的工程名、添加相机功能的目录和文件，并添加树莓派相机需要用到的库文件。按照下文提示找到相应的位置进行修改即可。

（1）设置本项目的工程名为 pro_9_realtime_video。

```
# -------------------------设置工程名-----------------------------
project(pro_9_realtime_video C)
# ---------------------------------------------------------------
```

（2）添加相机功能的目录、文件和库文件。

```
# step 1 添加相机功能源文件 camera_emu/*.c 和目录 ./camera_emu
file(GLOB_RECURSE MODULE_SAMPLE_SRC camera_emu/*.c)
include_directories(./camera_emu)

# step 2 添加树莓派头文件目录 ./raspi_camera/include
include_directories(./raspi_camera/include)

# step 3 添加连接目录 raspi_camera/lib
link_directories(raspi_camera/lib)
```

```
# step 4 添加树莓派连接库
${CMAKE_CURRENT_LIST_DIR}/raspi_camera/lib/libRaspiCamera.a
link_libraries(${CMAKE_CURRENT_LIST_DIR}/raspi_camera/lib/libRaspiCamera.a)
# ----------------------------------------------------------------------
```

5. 修改 main.c

（1）添加头文件。

```
/* --在此处添加相机功能头文件和相机媒体功能头文件--------------------------*/
#include "camera_emu/payload_cam_emu_media.h"
#include "camera_emu/payload_cam_emu_base.h"
/*--------------------------------------------------------------------*/
```

（2）添加相机功能和相机媒体功能的服务入口函数。

```
int main(int argc, char **argv)
{

    /* ----------在此处完成初始化相机功能和相机媒体功能-------------------------*/

    /* step 1 调用 DjiTest_CameraEmuBaseStartService 函数初始化相机功能-------*/
    returnCode = DjiTest_CameraEmuBaseStartService();
    if (returnCode != DJI_ERROR_SYSTEM_MODULE_CODE_SUCCESS) {
        USER_LOG_ERROR("camera emu common init error");
    }
    /* --------------------------------------------------------------------*/

    /* step 2 调用 DjiTest_CameraEmuMediaStartService 函数初始化相机媒体功能---*/
    returnCode = DjiTest_CameraEmuMediaStartService();
    if (returnCode != DJI_ERROR_SYSTEM_MODULE_CODE_SUCCESS) {
        USER_LOG_ERROR("camera emu media init error");
    }
    /* --------------------------------------------------------------------*/
}
```

6. 修改 payload_cam_emu_media.h

相机媒体功能头文件 payload_cam_emu_media.h 无须修改。

7. 修改 payload_cam_emu_media.c

根据 payload_cam_emu_media.c 的模板文件，需要完成以下两个函数的功能。

（1）完成服务入口函数。

在 DjiTest_CameraEmuMediaStartService 服务入口函数中完成以下功能。

```
/* -在 DjiTest_CameraEmuMediaStartService 服务入口函数中完成以下功能-------*/

T_DjiReturnCode DjiTest_CameraEmuMediaStartService(void)
{
```

```c
        T_DjiOsalHandler *osalHandler = DjiPlatform_GetOsalHandler();
        T_DjiReturnCode returnCode;

        /* step 1 定义高速数据传输功能的带宽比例为10%(数据流)、60%(视频流)和30%(下载流)*/
        const T_DjiDataChannelBandwidthProportionOfHighspeedChannel
bandwidthProportionOfHighspeedChannel = {10, 60, 30};
        /* -----------------------------------------------------------------------*/

/* step 2 调用DjiHighSpeedDataChannel_SetBandwidthProportion函数设置上述带宽比例*/
        returnCode = DjiHighSpeedDataChannel_SetBandwidthProportion
(bandwidthProportionOfHighspeedChannel);
        if (returnCode != DJI_ERROR_SYSTEM_MODULE_CODE_SUCCESS) {
            USER_LOG_ERROR("Set data channel bandwidth width proportion error.");
            return DJI_ERROR_SYSTEM_MODULE_CODE_UNKNOWN;
        }
        /* -----------------------------------------------------------------------*/

        if (aircraftInfoBaseInfo.aircraftType == DJI_AIRCRAFT_TYPE_M300_RTK) {

/* step 3 调用DjiPayloadCamera_RegMediaDownloadPlaybackHandler函数注册回放下载功能的函数-*/
            returnCode = DjiPayloadCamera_RegMediaDownloadPlaybackHandler
(&s_psdkCameraMedia);
            if (returnCode != DJI_ERROR_SYSTEM_MODULE_CODE_SUCCESS) {
                USER_LOG_ERROR("psdk camera media function init error.");
                return DJI_ERROR_SYSTEM_MODULE_CODE_UNKNOWN;
            }
        }
        /* -----------------------------------------------------------------------*/


/* step 4 创建名为UserCameraMedia_SendRaspiCamVideoTask的视频流处理线程----------*/
            returnCode = osalHandler->TaskCreate("user_camera_media_task",
UserCameraMedia_SendRaspiCamVideoTask, 2048,NULL, &s_userSendVideoThread);
            if (returnCode != DJI_ERROR_SYSTEM_MODULE_CODE_SUCCESS) {
                USER_LOG_ERROR("user send video task create error.");
                return DJI_ERROR_SYSTEM_MODULE_CODE_SUCCESS;
            }
        /* -----------------------------------------------------------------------*/
        }
```

（2）完成视频流处理线程。

在视频流处理线程UserCameraMedia_SendRaspiCamVideoTask中完成以下功能。

```c
    /* --在视频流处理线程UserCameraMedia_SendRaspiCamVideoTask中完成以下功能------- */

    static void *UserCameraMedia_SendRaspiCamVideoTask(void *arg)
    {
```

```c
    int ret;
    T_DjiReturnCode psdkStat;
    RASPI_VID_PARAMS camera_param;

    /* step 1 设置视频参数,分辨率为1280 × 720,每秒30帧--------------------*/
    camera_param.width = 1280;
    camera_param.height = 720;
    camera_param.framerate = 30;
    camera_param.addAud = 1;
    /* ------------------------------------------------------------------- */


    /* step 2 在启动实时视频流传输前,先停止可能存在的回放------------------*/
    psdkStat = DjiPlayback_StopPlayProcess();
    if (psdkStat != DJI_ERROR_SYSTEM_MODULE_CODE_SUCCESS) {
      USER_LOG_ERROR("stop playback and start liveview error: 0x%08llX.", psdkStat);
      //exit(1);
      return NULL;
    }
    /* ------------------------------------------------------------------- */


    /* step 3 设置H.264视频流接收函数(PsdkTest_RaspiCamH264Callback回调函数)-*/
    raspi_vid_callback_set(PsdkTest_RaspiCamH264Callback);
    /* ------------------------------------------------------------------- */


    /* step 4 调用树莓派main_process函数控制摄像头*/
    main_process(&camera_param);
    /* ------------------------------------------------------------------- */
}
```

8. 编译并执行项目

```
# 在pro_9_realtime_video目录下,创建build目录,并进入build目录
pi@raspberrypi:~/pro_9_realtime_video$ mkdir build && cd build

# 使用cmake创建项目编译环境
pi@raspberrypi:~/pro_9_realtime_video/build $ cmake ..

# make 编译
pi@raspberrypi:~/pro_9_realtime_video/build $ make

# 执行文件位于build/bin目录下
pi@raspberrypi:~/pro_9_realtime_video/build $cd bin/

# 使用sudo执行项目的执行文件
pi@raspberrypi:~/pro_9_realtime_video/build/bin $ sudo ./pro_9_realtime_video
```

任务评价

任务过程评价表

任务实施人姓名_____ 学号_____ 时间_____

评价项目及标准	分值	小组评议	教师评议
技术能力 1. 对基本概念的熟悉程度	10		
2. 掌握 H.264 的基本原理	10		
3. 掌握 PSDK 相机媒体功能	10		
4. 掌握连接功能库到项目的方法	10		
5. 掌握 PSDK 实现视频流的传输方法	10		
6. 掌握实时视频流传输与回放下载功能的测试方法	10		
执行能力 1. 出勤情况	5		
2. 遵守纪律情况	5		
3. 是否主动参与，有无提问记录	5		
4. 有无职业意识	5		
社会能力 1. 能否有效沟通	5		
2. 是否使用基本的文明礼貌用语	5		
3. 能否与组员主动交流、积极合作	5		
4. 能否自我学习、自我管理	5		
总分	100		
评定等级：			
评价意见		学习意见	

评定等级具体如下。

A：优，得分>90。B：好，90≥得分>80。C：一般，80≥得分>60。D：有待提高，得分≤60。

小结

本任务首先简要介绍了 H.264 视频编码技术；其次介绍了与实时视频流传输和回放下载相关的处理视频流、下载流的基本功能；最后介绍了实时视频流传输与回放下载功能的代码框架。

项目10 健康管理系统（HMS）

🔸 任务要求

健康管理系统（HMS）是一个对无人机各模块的运行健康状态进行监控的系统。DJI Pilot 提供了交互界面，该界面用于查看各模块的工作状态是否异常。本项目将通过 PSDK 提供的基本接口来获取各模块异常的错误信息，便于 PSDK 程序获知无人机各模块的工作状态。

🔸 知识导入

10.1 健康管理系统的基本概念

HMS 是监控无人机、相关传感器，挂载负载设备的系统。DJI Pilot 提供了交互界面，以查看各模块的工作状态是否异常。PSDK 端的 HMS 为机载设备端的开发提供了监控通道。通过对无人机及传感器状态的监控，针对性地调整负载设备功能，可以使负载设备更加安全和有效。

HMS 只从无人机端被动接收状态信息，所以在了解 HMS 功能前主要应了解无人机会给 PSDK 推送哪些健康信息。PSDK 3.x 版本中提供了更完善的无人机状态提示信息，因为提供的 HMS 信息较多，在此不再一一列举。具体请参考代码文件 dji_hms_info_table.h。

需要注意的是，HMS 是无人机端主动推送的，而 PSDK 端只能按照无人机端的定义格式解析，暂不支持自定义推送消息，即 dji_hms_info_table.h 中的信息不能被更改。

10.2 接口说明

HMS 相关功能的头文件为 dji_hms.h,该文件描述了 HMS 功能中结构体和函数原型的关键信息与使用方法。HMS 需要使用下面 3 个函数原型。
- DjiTest_HmsInit(void):初始化 HMS 模块。
- DjiTest_HmsDeInit(void):去初始化 HMS 模块。
- DjiHms_RegHmsInfoCallback(DjiHmsInfoCallback callback):注册回调函数,以获取 HMS 信息。

1. 初始化 HMS 模块

在使用 HMS 功能的相关接口之前,需要先调用 DjiTest_HmsInit 函数初始化 HMS 模块。

```
T_DjiReturnCode DjiTest_HmsInit(void);
```

2. 去初始化 HMS 模块

```
T_DjiReturnCode DjiTest_HmsDeInit(void);
```

3. 注册回调函数

将外部回调函数注册到 PSDK lib 中,注册的回调函数将以 1Hz 的频率被触发。

```
T_DjiReturnCode DjiHms_RegHmsInfoCallback(DjiHmsInfoCallback callback);
```

10.3 代码模板

1. 代码结构

下面是实现 HMS 功能的代码框架,用户只需要编写用于 HMS 功能的 hms.c 和 hms.h 即可。

```
.
├── 3rdparty
├── CMakeLists.txt
├── dji_init.c
├── dji_init.h
├── hal
├── hms
│   ├── hms.c
│   └── hms.h
├── main.c
├── monitor
```

```
├── osal
└── psdk_lib
```

2. 头文件模板

本项目的头文件名为 hms.h。该文件定义的内容可满足正常的应用开发需求，如果需要实现特殊的功能，则可以按照需要进行修改。

这里不再展示头文件代码。

3. 源文件模板

本项目的源文件名为 hms.c，以下是 hms.c 的模板文件，按照提示即可完成代码的编写。

由于源文件的代码行数太多，这里仅给出在 hms.c 中需要完成的功能。

（1）在服务入口函数中完成以下功能。

```c
/*-----------在服务入口函数 DjiTest_HmsRunSample 中完成以下功能------*/
/* step 1 调用 DjiTest_HmsInit 函数初始化 HMS 模块--------------------*/
/* step 2 调用 DjiHms_RegHmsInfoCallback 函数注册 DjiTest_HmsInfoCallback 回调函数-----*/

    T_DjiReturnCode DjiTest_HmsRunSample(void)
    {
        T_DjiReturnCode returnCode;
        T_DjiOsalHandler *osalHandler;

        USER_LOG_INFO("Hms Sample Start");
        USER_LOG_INFO("--> Step 1: Init hms sample");

        /* step 1 调用 DjiTest_HmsInit 函数初始化 HMS 模块--------------------*/

        /* ----------------------------------------------------------------*/
        osalHandler = DjiPlatform_GetOsalHandler();
        USER_LOG_INFO("--> Step 2: Register callback function of push HMS information");

/* step 2 调用 DjiHms_RegHmsInfoCallback 函数注册 DjiTest_HmsInfoCallback 回调函数-----*/

        /* ----------------------------------------------------------------*/
        printf("register success------------1\n");
        osalHandler->TaskSleepMs(10000);
    out:
        USER_LOG_INFO("--> Step 3: Deinit hms sample");
        returnCode = DjiTest_HmsDeInit();
        if (returnCode != DJI_ERROR_SYSTEM_MODULE_CODE_SUCCESS) {
            USER_LOG_ERROR("Hms sample deinit error, error code:0x%08llX", returnCode);
        }
        USER_LOG_INFO("Hms Sample End");
        return returnCode;
    }
```

（2）在 HMS 初始化函数中完成以下功能。

```
/*-----------在初始化函数 DjiTest_HmsInit 中完成以下功能------------------*/
/*step 1 调用 DjiFcSubscription_Init 函数初始化消息订阅功能------------*/
/*step 2 调用 DjiFcSubscription_SubscribeTopic 函数订阅无人机状态数据-----*/
static T_DjiReturnCode DjiTest_HmsInit(void)
{
    T_DjiReturnCode returnCode;

    /*step 1 调用 DjiFcSubscription_Init 函数初始化消息订阅功能------------*/

    /* ----------------------------------------------------------------*/

    /*step 2 调用 DjiFcSubscription_SubscribeTopic 函数订阅无人机状态数据-----*/

    /* ----------------------------------------------------------------*/
    USER_LOG_INFO(" DjiHms_Init() init HMS module");
    return DjiHms_Init();
}
```

（3）在获取数据函数中完成以下功能。

```
/*-----------在 DjiTest_GetValueOfFlightStatus 函数中完成以下功能------------*/
/*step 调用 DjiFcSubscription_GetLatestValueOfTopic 函数获取无人机数据------*/
static T_DjiFcSubscriptionFlightStatus DjiTest_GetValueOfFlightStatus(void)
{
    T_DjiReturnCode djiStat;
    T_DjiFcSubscriptionFlightStatus flightStatus;
    T_DjiDataTimestamp flightStatusTimestamp = {0};

    /*step 调用 DjiFcSubscription_GetLatestValueOfTopic 函数获取无人机数据--*/

    /* ----------------------------------------------------------------*/
    if (djiStat != DJI_ERROR_SYSTEM_MODULE_CODE_SUCCESS) {
        USER_LOG_ERROR("Get value of topic flight status failed, error code:0x%08llX", djiStat);
        flightStatus = 0;
    } else {
        USER_LOG_DEBUG("Timestamp: millisecond %u microsecond %u.",
flightStatusTimestamp.millisecond,
                        flightStatusTimestamp.microsecond);
        USER_LOG_DEBUG("Flight status: %d.", flightStatus);
    }
    return flightStatus;
}
```

任务设计

本项目将通过 PSDK3.x 提供 HMS 获取无人机各模块异常的错误信息，从而获取无人机各模块的工作状态。

整个任务包括两个部分，一个是初始化 HMS 模块（注意 return 部分），另一个是初始化消息订阅模块。在 HMS 中之所以需要消息订阅模块，是因为在后续获取状态解析时涉及无人机是在空中还是在地面。

任务的具体设计要求如下。

（1）创建项目目录。

（2）本项目将实现 HMS 功能，项目目录下的 hms 目录用于编写代码，以实现用户线程调用 HMS API 接口。

（3）在 hms 目录下编写代码，实现 HMS 功能。

① 初始化 HMS 模块。

② 初始化消息订阅模块。

③ 订阅数据项。

④ 获取订阅的数据项。

（4）使用 cmake 编译项目。

（5）执行项目。

任务实施

限于篇幅，这里只给出任务实施步骤和关键代码。

1. 模板文件

连接树莓派开发板，默认路径为当前登录用户的家目录。查看当前的家目录，该目录下有一个名为 pro_10_hms_source_template 的目录，此目录下保存了本项目的所有模板文件。

目录结构如下。

```
.
├── 3rdparty
├── CMakeLists.txt
├── dji_init.c
├── dji_init.h
├── hal
├── hms
│   ├── hms.c
│   └── hms.h
├── main.c
├── monitor
├── osal
└── psdk_lib
```

2. 复制模板目录

将 pro_10_hms _source_template 目录下的所有文件和目录复制到 pro_10_hms 目录中。

```
pi@raspberrypi:~ $ cp -r pro_10_hms _source_template pro_10_hms
```

3. 修改 dji_init.c

填写用户账号信息。

```
#define USER_App_NAME                "RaspberryPi"
#define USER_App_ID                  "126413"
#define USER_App_KEY                 "0ea2eadb314d1a6e870975364e2bf4a"
#define USER_App_LICENSE             "HZ/GptycWQ........................"
#define USER_DEVELOPER_ACCOUNT       "abc@outlo**.com"
#define USER_BAUD_RATE               "460800"
```

4. 修改 CMakeLists.txt

（1）设置本项目的工程名为 pro_10_hms。

```
# --------------------设置工程名------------------------------
project(pro_10_hms C)
# ----------------------------------------------------------
```

（2）添加目录、文件。

```
# -------在此处添加 HMS 的目录及该目录下的所有.c 文件-----------------
file(GLOB_RECURSE MODULE_SAMPLE_SRC hms/*.c)
include_directories(./hms)
# ----------------------------------------------------------
```

5. 修改 main.c

（1）添加头文件。

```
/*----------在此处添加 HMS 头文件-----------------------*/
#include <hms/hms.h>
/* ---------------------------------------------------*/
```

（2）添加 HMS 功能服务入口函数。

```
int main(int argc, char **argv)
{
    /*----------在此处添加初始化 HMS 功能的代码----------------------*/
    returnCode = DjiTest_HmsRunSample();
    if (returnCode != DJI_ERROR_SYSTEM_MODULE_CODE_SUCCESS) {
        USER_LOG_ERROR("HMS init error\n");
    }
    /*----------------------------------------------------------*/
}
```

6. 修改 hms.h

HMS 头文件 hms.h 无须修改。

7. 修改 hms.c

根据 hms.c 的模板文件，需要完成以下 3 个函数的功能。

（1）完成服务入口函数。

在服务入口函数 DjiTest_HmsRunSample 中完成以下功能。

```
/*----------在服务入口函数 DjiTest_HmsRunSample 中完成以下功能---------*/

T_DjiReturnCode DjiTest_HmsRunSample(void)
{
    /* step 1 调用 DjiTest_HmsInit 函数完成 HMS 模块的初始化--------------------*/
    returnCode = DjiTest_HmsInit();
    if (returnCode != DJI_ERROR_SYSTEM_MODULE_CODE_SUCCESS) {
        USER_LOG_ERROR("Hms sample init error, error code:0x%08llX", returnCode);
        goto out;
    }
    /* ------------------------------------------------------------------*/

    /* step 2 调用 DjiHms_RegHmsInfoCallback 函数注册 DjiTest_HmsInfoCallback 回调函数-----*/
    returnCode = DjiHms_RegHmsInfoCallback(DjiTest_HmsInfoCallback);
    if (returnCode != DJI_ERROR_SYSTEM_MODULE_CODE_SUCCESS) {
        USER_LOG_ERROR("Register callback function of push HMS information failed, error code:0x%08llX", returnCode);
        goto out;
    }
    /* ------------------------------------------------------------------*/
}
```

（2）完成 HMS 初始化函数。

```
/*----------在初始化函数 DjiTest_HmsInit 中完成以下功能-----------------*/

static T_DjiReturnCode DjiTest_HmsInit(void)
{
    T_DjiReturnCode returnCode;

    /*step 1 调用 DjiFcSubscription_Init 函数初始化消息订阅模块-----------*/
    returnCode = DjiFcSubscription_Init();
    USER_LOG_INFO(" init data subscription module");
    if (returnCode != DJI_ERROR_SYSTEM_MODULE_CODE_SUCCESS) {
        USER_LOG_ERROR("Hms sample init data subscription module failed, error code:0x%08llX", returnCode);
        return returnCode;
    }
    /* ------------------------------------------------------------------*/

    /*step 2 调用 DjiFcSubscription_SubscribeTopic 函数订阅无人机状态数据-----*/
    returnCode = DjiFcSubscription_SubscribeTopic(
    DJI_FC_SUBSCRIPTION_TOPIC_STATUS_FLIGHT,
```

```
                    DJI_DATA_SUBSCRIPTION_TOPIC_10_HZ,
                    NULL);
    USER_LOG_INFO(" SubscribeTopic");
    if (returnCode != DJI_ERROR_SYSTEM_MODULE_CODE_SUCCESS) {
        USER_LOG_ERROR("HMS sample subscribe topic flight status error, error
code:0x%08llX", returnCode);
        return returnCode;
    }
    /* ----------------------------------------------------------------*/
    return DjiHms_Init();
}
```

（3）完成获取无人机数据的函数。

```
static T_DjiFcSubscriptionFlightStatus DjiTest_GetValueOfFlightStatus(void)
{
    T_DjiReturnCode djiStat;
    T_DjiFcSubscriptionFlightStatus flightStatus;
    T_DjiDataTimestamp flightStatusTimestamp = {0};

    /*step 调用 DjiFcSubscription_GetLatestValueOfTopic 函数获取无人机数据--*/
    djiStat = DjiFcSubscription_GetLatestValueOfTopic
                        (DJI_FC_SUBSCRIPTION_TOPIC_STATUS_FLIGHT,
                                    (uint8_t *) &flightStatus,
                            sizeof(T_DjiFcSubscriptionFlightStatus),
                                        &flightStatusTimestamp);
    printf("Flight status: %d.", flightStatus);
    /* ----------------------------------------------------------------*/

}
```

8. 编译并执行项目

```
# 在 pro_10_hms 目录下，创建 build 目录，并进入 build 目录
pi@raspberrypi:~/pro_10_hms$ mkdir build && cd build

# 使用 cmake 创建项目编译环境
pi@raspberrypi:~/pro_10_hms/build $ cmake ..

# make 编译
pi@raspberrypi:~/pro_10_hms/build $ make

# 执行文件位于 build/bin 目录下
pi@raspberrypi:~/pro_10_hms/build $cd bin/

# 使用 sudo 执行项目的执行文件
pi@raspberrypi:~/pro_10_hms/build/bin $ sudo ./pro_10_hms
```

任务评价

任务过程评价表

任务实施人姓名_____ 学号_____ 时间_____

评价项目及标准		分值	小组评议	教师评议
技术能力	1. 对基本概念的熟悉程度	10		
	2. 掌握 HMS 的作用	10		
	3. 掌握订阅无人机状态数据的方法	10		
	4. 掌握获取无人机状态数据的方法	10		
	5. 掌握构建项目的方法	10		
	6. 了解 HMS 部分信息码	10		
执行能力	1. 出勤情况	5		
	2. 遵守纪律情况	5		
	3. 是否主动参与，有无提问记录	5		
	4. 有无职业意识	5		
社会能力	1. 能否有效沟通	5		
	2. 是否使用基本的文明礼貌用语	5		
	3. 能否与组员主动交流、积极合作	5		
	4. 能否自我学习、自我管理	5		
总分		100		
评定等级：				
评价意见			学习意见	

评定等级具体如下。

A：优，得分>90。B：好，90≥得分>80。C：一般，80≥得分>60。D：有待提高，得分≤60。

小结

本项目首先简要介绍了健康管理系统的基本概念；其次介绍了与健康管理系统相关的接口；最后介绍了健康管理系统的代码框架。

项目11

获取相机码流（liveview）

◎ 任务要求

本项目将通过实时视频流功能获取无人机机载摄像头的相机码流。

◎ 知识导入

11.1 实时视频流

实时视频流的传输需要一定的带宽能力，且必须使用 USB。机载计算机通过此功能可以实时获取负载相机或 FPV 相机的视频数据，还可以结合实际应用进行视频图像处理，并对视频流进行转发等。实时视频流功能仅支持 Linux 平台，RTOS 不支持读取实时视频流。

为了满足开发者使用基于 PSDK 开发的应用程序获取相机码流的需求，PSDK 提供了获取相机码流的功能，支持获取 FPV 相机或获取 1 号云台相机的 H.264 码流和 RGB 图像。

M300 RTK 无人机支持获取 FPV 相机和所有云台相机的 H.264 码流。由于获取 FPV 相机码流和获取 1 号云台上的相机码流的回调函数在各自独立的线程中运行，且 OpenCV 的 imshow 模块仅支持在一个线程中运行，因此仅支持开发者获取 FPV 相机或获取 1 号云台相机的 H.264 码流和 RGB 图像。获取相机码流后，使用 FFmpeg 等解码器解码。

11.2 接口说明

获取相机码流相关功能的头文件为 dji_liveview.h，该文件描述了相机码流功能中结构体和函数原型的关键信息与使用方法。相机管理函数如表 11-1 所示。

表 11-1 相机管理函数

函数	功能
DjiLiveview_Init	liveview 模块初始化
DjiLiveview_Deinit	liveview 模块去初始化
DjiLiveview_StartH264Stream	按照选定位置启动 FPV 或摄像机 H.264 码流
DjiLiveview_StopH264Stream	按照选定位置关闭 FPV 或摄像机 H.264 码流

1. liveview 模块初始化

```
T_DjiReturnCode DjiLiveview_Init(void);
```

2. liveview 模块去初始化

```
T_DjiReturnCode DjiLiveview_Deinit(void);
```

3. 按照选定位置启动 FPV 或摄像机 H.264 码流

```
T_DjiReturnCode DjiLiveview_StartH264Stream(
            E_DjiLiveViewCameraPosition position,
            E_DjiLiveViewCameraSource source,
            DjiLiveview_H264Callback callback);
```

说明：参数 position 指出哪个相机输出 H.264 码流，参数 source 指向摄像头输出 H.264 码流，参数 callback 是当接收到新的 H.264 帧时，在回调线程中调用的回调函数。

4. 按照选定位置关闭 FPV 或摄像机 H.264 码流

```
T_DjiReturnCode DjiLiveview_StopH264Stream(
            E_DjiLiveViewCameraPosition position);
```

说明：参数 position 指出哪个相机输出 H.264 码流。

11.3 代码模板

1. 代码结构

下面是实现 liveview 功能的代码结构，用户只需要编写用于实现 liveview 功能的 liveview.c 和 liveview.h 即可。

```
.
├── 3rdparty
├── CMakeLists.txt
├── dji_init.c
├── dji_init.h
├── hal
├── liveview
│   ├── liveview.c
│   └── liveview.h
├── main.c
├── monitor
├── osal
└── psdk_lib
```

2. 头文件模板

本项目的头文件名为 liveview.h。该文件定义的内容可满足正常的应用开发需求，如果需要实现特殊的功能，则可以按照需要进行修改。

这里不再展示头文件代码。

3. 源文件模板

本项目的源文件名为 liveview.c，以下是 liveview.c 的模板文件，按照提示即可完成代码的编写。

由于源文件的代码行数太多，这里仅给出在 liveview.c 中需要完成的功能。

在服务入口函数中完成以下功能。

```c
/*----在服务入口函数中完成以下功能--------------------------------*/
/*step 1 初始化 liveview 功能*/
/*step 2 启动 FPV H.264 码流，回调函数为 DjiTest_FpvCameraStreamCallback*/
/*step 3 关闭 FPV H.264 码流 ---------------------------------*/

T_DjiReturnCode DjiTest_LiveviewRunSample(E_DjiMountPosition mountPosition)
{
    T_DjiReturnCode returnCode;
    T_DjiOsalHandler *osalHandler = DjiPlatform_GetOsalHandler();
    time_t currentTime = time(NULL);
```

```c
            struct tm *localTime = NULL;
            T_DjiAircraftInfoBaseInfo aircraftInfoBaseInfo = {0};

            USER_LOG_INFO("Liveview sample start");

            returnCode = DjiAircraftInfo_GetBaseInfo(&aircraftInfoBaseInfo);
            if (returnCode != DJI_ERROR_SYSTEM_MODULE_CODE_SUCCESS) {
                USER_LOG_ERROR("get aircraft base info error");
                return DJI_ERROR_SYSTEM_MODULE_CODE_SYSTEM_ERROR;
            }

            USER_LOG_INFO("--> Step 1: Init liveview module");

            /*step 1 初始化liveview功能--------------------------------*/

            /*-------------------------------------------------------*/

            USER_LOG_INFO("--> Step 2: Start h264 stream of the fpv\r\n");

            localTime = localtime(&currentTime);
            sprintf(s_fpvCameraStreamFilePath, "fpv_stream_%04d%02d%02d_%02d-%02d-%02d.h264",
                    localTime->tm_year + 1900, localTime->tm_mon + 1, localTime->tm_mday,
                    localTime->tm_hour, localTime->tm_min, localTime->tm_sec);

            /*step 2 启动FPV H.264码流，回调函数为DjiTest_FpvCameraStreamCallback-*/

            /* ------------------------------------------------------------*/

            /* ----------获取20秒的H.264视频流-----------------------*/
            for (int i = 0; i < TEST_LIVEVIEW_STREAM_STROING_TIME_IN_SECONDS; ++i) {
                USER_LOG_INFO("Storing camera h264 stream, second: %d.", i + 1);
                osalHandler->TaskSleepMs(1000);
            }

            USER_LOG_INFO("--> Step 3: Stop h264 stream of the fpv and selected payload\r\n");

            /*step 3 关闭FPV H.264码流 ----------------------------------*/

            /* ------------------------------------------------------------*/
            USER_LOG_INFO("Fpv stream is saved to file: %s", s_fpvCameraStreamFilePath);

            USER_LOG_INFO("--> Step 4: Deinit liveview module");
            returnCode = DjiLiveview_Deinit();
            if (returnCode != DJI_ERROR_SYSTEM_MODULE_CODE_SUCCESS) {
                USER_LOG_ERROR("Liveview deinit failed, error code: 0x%08X", returnCode);
```

```
        goto out;
    }

out:
    USER_LOG_INFO("Liveview sample end");

    return returnCode;
}
```

任务设计

本项目将通过 PSDK3.x 提供的 liveview 功能获取无人机 FPV 相机的实时视频流。
任务的具体设计要求如下。
（1）创建项目目录。
（2）本项目将实现 liveview 功能，项目目录下的 liveview 目录用于编写代码，以实现用户线程调用 liveview API 接口。
（3）在 liveview 目录下编写代码，实现 liveview 功能。
① 初始化 liveview 模块。
② 查看树莓派中的 USB 接口。
③ 通过 FPV 相机进行视频流传输。
（4）使用 cmake 编译项目。
（5）执行项目。

任务实施

限于篇幅，这里只给出任务实施步骤和关键代码。

1. 模板文件

连接树莓派开发板，默认路径为当前登录用户的家目录。查看当前的家目录，该目录下有一个名为 pro_11_liveview _source_template 的目录，此目录下保存了本项目的所有模板文件。

目录结构如下。

```
.
├── 3rdparty
├── CMakeLists.txt
├── dji_init.c
├── dji_init.h
├── hal
├── liveview
│   ├── liveview.c
```

```
    |    └── liveview.h
    ├── main.c
    ├── monitor
    ├── osal
    └── psdk_lib
```

2. 复制模板目录

将 pro_11_liveview_source_template 目录下的所有文件和目录复制到 pro_11_liveview 目录中。

```
pi@raspberrypi:~ $ cp -r pro_11_liveview_source_template pro_11_liveview
```

3. 修改 dji_init.c

填写用户账号信息。

```
#define USER_App_NAME              "RaspberryPi"
#define USER_App_ID                "126413"
#define USER_App_KEY               "0ea2eadb314d1a6e870975364e2bf4a"
#define USER_App_LICENSE           "HZ/GptycWQ........................"
#define USER_DEVELOPER_ACCOUNT     "abc@outlo**.com"
#define USER_BAUD_RATE             "460800"
```

4. 修改 CMakeLists.txt

（1）设置本项目的工程名为 pro_11_liveview。

```
# --------------设置工程名--------------------------------------
project(pro_11_liveview C)
# ------------------------------------------------------------
```

（2）添加目录、文件。

```
# -------在此处添加 liveview 的目录及该目录下的所有 .c 文件--------------------
file(GLOB_RECURSE MODULE_SAMPLE_SRC liveview/*.c)
include_directories(./liveview)
# ------------------------------------------------------------
```

5. 在 hal/hal_uart.h 中设置 USB 设备

（1）在树莓派上查看 USB 设备。

```
pi@raspberrypi:~$ ls -l /dev/ttyA*
crw-rw---- 1 root dialout 166,  0 Mar 18 10:16  /dev/ttyACM0
crw-rw---- 1 root tty     204, 64 Mar 11 17:31  /dev/ttyAMA0
pi@raspberrypi:~$ ls -l /dev/ttyUSB*
crwxrwxrwx 1 root dialout 188,  0 Mar 17 19:59  /dev/ttyUSB0
```

（2）在 hal_uart.h 中定义 USB 设备。

```
/*-----在 hal_uart.h 中定义 USB 设备--------------------------*/
#define LINUX_UART_DEV1    "/dev/ttyUSB0"
#define LINUX_UART_DEV2    "/dev/ttyACM0"
/*----------------------------------------------------------*/
```

6. 修改 main.c

（1）添加头文件。

```
/*----------在此处添加 liveview.h 和 dji_typedef.h----------*/
#include <liveview/liveview.h>
#include "dji_typedef.h"
/*---------------------------------------------------------*/
```

（2）设置 FPV 相机位置。

按照枚举类型 E_DjiLiveViewCameraPosition 设置 FPV 相机位置为 7。

```
/* -----在此设置 FPV 相机位置 DJI_LIVEVIEW_CAMERA_POSITION_FPV = 7----*/
static int s_mountPosition = 7;
```

（3）在 main 函数中添加 liveview 服务入口函数。

```
int main(int argc, char **argv)
{
    /* ---调用 liveview 服务入口函数--------------------------*/
    returnCode = DjiTest_LiveviewRunSample(s_mountPosition);
    if (returnCode != DJI_ERROR_SYSTEM_MODULE_CODE_SUCCESS) {
        USER_LOG_ERROR("LiveView init error\n");
    }
    /* ----------------------------------------------------*/
}
```

7. 修改 liveview.h

liveview.h 无须修改。

8. 修改 liveview.c

在服务入口函数 DjiTest_LiveviewRunSample 中完成以下功能。

```
/*----在服务入口函数中完成以下功能----------------------------*/

T_DjiReturnCode DjiTest_LiveviewRunSample(E_DjiMountPosition mountPosition)
{

    /*step 1 初始化 liveview 模块--------------------------*/
    returnCode = DjiLiveview_Init();
    if (returnCode != DJI_ERROR_SYSTEM_MODULE_CODE_SUCCESS) {
        USER_LOG_ERROR("Liveview init failed, error code: 0x%08X", returnCode);
        goto out;
    }
    /*---------------------------------------------------*/

    /*step 2 启动 FPV H.264 码流，回调函数为 DjiTest_FpvCameraStreamCallback*/
    returnCode = DjiLiveview_StartH264Stream(DJI_LIVEVIEW_CAMERA_POSITION_FPV,
        DJI_LIVEVIEW_CAMERA_SOURCE_DEFAULT,
            DjiTest_FpvCameraStreamCallback);
```

```c
        if (returnCode != DJI_ERROR_SYSTEM_MODULE_CODE_SUCCESS) {
            USER_LOG_ERROR("Request h264 of fpv failed, error code: 0x%08X", returnCode);
            goto out;
        }
        /* ----------------------------------------------------------*/


        /* ----------获取20秒的H.264视频流----------------------*/
        for (int i = 0; i < TEST_LIVEVIEW_STREAM_STROING_TIME_IN_SECONDS; ++i) {
            USER_LOG_INFO("Storing camera h264 stream, second: %d.", i + 1);
            osalHandler->TaskSleepMs(1000);
        }

    USER_LOG_INFO("--> Step 3: Stop h264 stream of the fpv and selected payload\r\n");

        /*step 3 关闭 FPV H.264 码流 ---------------------------------------*/
        returnCode = DjiLiveview_StopH264Stream(
    DJI_LIVEVIEW_CAMERA_POSITION_FPV,
    DJI_LIVEVIEW_CAMERA_SOURCE_DEFAULT);
        if (returnCode != DJI_ERROR_SYSTEM_MODULE_CODE_SUCCESS) {
            USER_LOG_ERROR("Request to stop h264 of fpv failed, error code: 0x%08X", returnCode);
            goto out;
        }
        /* ----------------------------------------------------------*/


    }
```

9. 编译并执行项目

```
# 在pro_11_liveview目录下，创建build目录，并进入build目录
pi@raspberrypi:~/pro_11_liveview$ mkdir build && cd build

# 使用cmake创建项目编译环境
pi@raspberrypi:~/pro_11_liveview/build $ cmake ..

# make 编译
pi@raspberrypi:~/pro_11_liveview/build $ make

# 执行文件位于build/bin目录下
pi@raspberrypi:~/pro_11_liveview/build $cd bin/

# 使用sudo执行项目的执行文件
pi@raspberrypi:~/pro_11_liveview/build/bin $ sudo ./pro_11_liveview
```

任务评价

任务过程评价表

任务实施人姓名_____ 学号_____ 时间_____

	评价项目及标准	分值	小组评议	教师评议
技术能力	1. 对基本概念的熟悉程度	10		
	2. 掌握 liveview 的通信方式	10		
	3. 掌握无人机 FPV 相机	10		
	4. 了解 liveview 处理 H.264 码流的基本方法	10		
	5. 掌握 liveview 运行的基本流程	10		
	6. 掌握 liveview 的测试方法	10		
执行能力	1. 出勤情况	5		
	2. 遵守纪律情况	5		
	3. 是否主动参与，有无提问记录	5		
	4. 有无职业意识	5		
社会能力	1. 能否有效沟通	5		
	2. 是否使用基本的文明礼貌用语	5		
	3. 能否与组员主动交流、积极合作	5		
	4. 能否自我学习、自我管理	5		
总分		100		
评定等级：				
评价意见		学习意见		

评定等级具体如下。

A：优，得分>90。B：好，90≥得分>80。C：一般，80≥得分>60。D：有待提高，得分≤60。

小结

本项目首先简要介绍了 PSDK3.x 获取相机码流的相关知识；其次介绍了与获取相机码流相关的接口说明；最后介绍了获取相机码流的代码框架。

项目12

本地升级

⊙ 任务要求

本项目在已完成的"项目 9 任务 1 视频流文件传输与回放下载"的基础上,通过 PSDK 的本地升级功能,将负载设备的固件从版本 01.00.00.00 升级到版本 01.00.00.01。

⊙ 知识导入

12.1 本地升级功能

本地升级是用于升级负载设备固件的功能,开发者应用本地升级功能能够使用户通过 DJI Assistant 2 方便地选择本地固件文件并更新负载设备的固件。

将基于 PSDK 发布的负载设备交给用户后,若后续版本升级,则开发者应该将新版本的程序固件包发给用户,用户通过 DJI Assistant 2 导入程序固件包更新负载设备,进行本地升级。

在开发本地升级功能时,可调用 DjiUpgrade_Init 函数指定传输协议并配置传输设置信息。

```
returnCode = DjiUpgrade_Init(&upgradeConfig);
if (returnCode != DJI_ERROR_SYSTEM_MODULE_CODE_SUCCESS) {
    USER_LOG_ERROR("DjiUpgrade_Init error, return code = %d", returnCode);
    return returnCode;
}
```

本地升级支持 FTP、DCFTP 两种传输协议,在 dji_upgrade.h 中定义了枚举类型 E_DjiFirmwareTransferType。

```
typedef enum {
    DJI_FIRMWARE_TRANSFER_TYPE_FTP = 0,
    DJI_FIRMWARE_TRANSFER_TYPE_DCFTP,
} E_DjiFirmwareTransferType;
```

1. 使用 FTP

PSDK 支持开发者使用 FTP 传输所需升级的固件，在开发者使用 FTP 传输固件时，应为负载设备部署相应的 FTP 服务器。需要注意以下内容。

（1）只有基于 Linux 平台开发的负载设备才支持通过网口使用 FTP 传输所需升级的固件。

（2）使用 FTP 传输负载设备固件时的用户账号为 psdk_payload_ftp，密码为 DJi_#$31。

（3）使用 FTP 传输负载设备固件时的负载设备为 FTP 服务器，无人机为 FTP 客户端。

（4）升级的固件文件先通过 DJI Assistant 2（串口）传给无人机，无人机再通过 FTP 传输到负载设备中。

本地升级拓扑如图 12-1 所示。

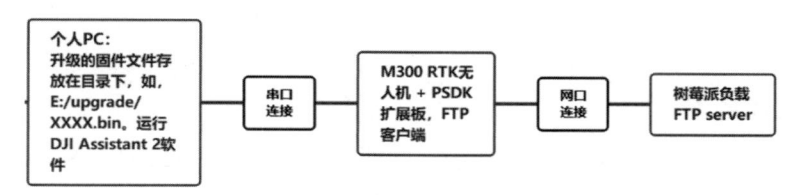

图 12-1 本地升级拓扑

2. 使用 DCFTP

DCFTP（DJI Common File Transfer Protocol）是 DJI 独有的文件传输协议，开发者通过调用指定的接口即可获得所需的文件。需要注意的是，在使用 RTOS 或没有网口的负载设备时，可使用该方式传输固件。由于该方式使用命令信号传输通道传输固件，因此传输速度较慢。

12.2 固件版本

在升级负载设备固件时，版本不可低于当前版本，即当负载设备当前使用的固件版本号为 01.00.00.00 时，升级的版本号可以改为 01.00.00.01。

1. 固件版本号

在 dji_upgrade.h 中定义了表示设置固件版本的结构体 T_DjiFirmwareVersion。

```
typedef struct {
    uint8_t majorVersion; /* major 固件版本，范围为 0 ～ 99 */
    uint8_t minorVersion; /* minor 固件版本，范围为 0 ～ 99 */
    uint8_t modifyVersion; /* modify 固件版本，范围为 0 ～ 99 */
    uint8_t debugVersion; /* debug 固件版本，范围为 0 ～ 99 */
} T_DjiFirmwareVersion;
```

在 dji_upgrade.h 中定义了枚举类型 E_DjiFirmwareTransferType。使用 FTP 传输时的用户名和密码分别为 psdk_payload_ftp 和 DJi_#$31。

```
typedef enum {
    DJI_FIRMWARE_TRANSFER_TYPE_FTP = 0,  /* FTP 固件传输类型*/
```

```
        DJI_FIRMWARE_TRANSFER_TYPE_DCFTP,         /* DCFTP 固件传输类型*/
} E_DjiFirmwareTransferType;
```

在 dji_upgrade.h 中定义了结构体 T_DjiUpgradeConfig，用于指定升级的版本和传输协议。

```
typedef struct {
    T_DjiFirmwareVersion currentFirmwareVersion;      /*当前固件版本 */
    T_DjiFirmwareTransferInfo firmwareTransferInfo;  /* 传输协议信息 */
} T_DjiUpgradeConfig;
```

在 main.c 中定义升级的固件版本和传输协议。

```
T_DjiTestUpgradeConfig testUpgradeConfig = {
    .firmwareVersion = {1, 0, 0, 0},     /* 定义固件版本 */
    .transferType = DJI_FIRMWARE_TRANSFER_TYPE_FTP, /* 定义传输协议 */
    .needReplaceProgramBeforeReboot = true
};
```

2. 固件文件名

新编译固件文件后，需修改编译后的固件文件名，如 PSDK_AppALIAS_V02.01.00.04.bin。文件名格式为 Product Name_Vaa.bb.cc.dd.bin。其中：

- **Product Name**：负载产品名（若开发者设置了负载别名，则该名称为负载别名）。
- **Vaa.bb.cc.dd**：为负载 4 位版本号。

需要注意的是，固件文件名中的版本号要与代码中定义的版本号一致，否则负载设备可能会升级失败。

3. 使用 DJI Assistant 2 升级固件

打开 DJI Assistant 2，选择编译生成的固件升级文件（PSDK_AppALIAS_V02.01.00.04.bin）进行固件升级。使用 DJI Assistant 2 升级负载设备固件如图 12-2 所示。

图 12-2　使用 DJI Assistant 2 升级负载设备固件

12.3 接口说明

本节主要介绍本地升级涉及的头文件 dji_upgrade.h 和 upgrade_platform_opt_linux.h。

12.3.1 本地升级模块

本地升级相关功能的头文件为 dji_upgrade.h，该文件描述了本地升级功能中结构体和函数原型的关键信息与使用方法。目录结构如下。

```
pi@raspberrypi:~/pro_12_upgrade/upgrade $ tree
.
├── upgrade.c
├── upgrade_common_file_transfer.c
├── upgrade_common_file_transfer.h
├── upgrade.h
├── upgrade_platform_opt.c
└── upgrade_platform_opt.h

0 directories, 6 files
```

在 dji_upgrade.h 中定义了以下宏定义、枚举、结构体和函数原型。

```
# 宏定义、枚举与结构体
E_DjiFirmwareTransferType
E_DjiUpgradeStage
E_DjiUpgradeEndState
T_DjiFirmwareVersion
T_DjiUpgradeOngoingInfo
T_DjiUpgradeRebootInfo
T_DjiUpgradeEndInfo
T_DjiUpgradeFtpFileTransferInfo
T_DjiUpgradeFileInfo
T_DjiUpgradeDcftpFileTransferOpt
T_DjiFirmwareTransferInfo
T_DjiUpgradeState
T_DjiUpgradeConfig
T_DjiUpgradeHandler

# 函数原型
DjiUpgrade_Init                    /*本地升级模块初始化*/
DjiUpgrade_EnableLocalUpgrade      /*启动本地升级模块*/
DjiUpgrade_RegHandler              /*注册升级过程的处理程序*/
DjiUpgrade_PushUpgradeState        /*向终端 App 推送升级状态*/
```

12.3.2 升级操作

升级操作包含在 upgrade_platform_opt 目录下的 upgrade_platform_opt_linux.c 和 upgrade_platform_opt_linux.h 中。目录结构如下。

```
pi@raspberrypi:~/pro_12_upgrade/upgrade_platform_opt $ tree
.
├── upgrade_platform_opt_linux.c
└── upgrade_platform_opt_linux.h

0 directories, 2 files
```

upgrade_platform_opt_linux.h 中定义了升级时老版本的路径和新版本下载后存放的目录，以及升级的各种操作函数。说明如下。

```
/* Define to prevent recursive inclusion ---------------------------*/
#ifndef UPGRADE_PLATFORM_OPT_LINUX_H
#define UPGRADE_PLATFORM_OPT_LINUX_H

/* Includes --------------------------------------------------------*/
#include <dji_typedef.h>
#include <dji_upgrade.h>

#ifdef __cplusplus
extern "C" {
#endif

/* Exported constants ----------------------------------------------*/
/* 老版本固件运行路径，默认为"/usr/local/bin/dji_sdk_demo_linux"，可修改 */
#define DJI_TEST_UPGRADE_OLD_FIRMWARE_PATH   "/usr/local/bin/dji_sdk_demo_linux"

/* 通过FTP下载的新版本固件存放目录，默认为"/upgrade/"，可修改 */
#define DJI_TEST_UPGRADE_FILE_DIR         "/upgrade/"

/* Exported types --------------------------------------------------*/
/* 升级后重启系统 */
T_DjiReturnCode DjiUpgradePlatformLinux_RebootSystem(void);

/* 升级后清除缓存 */
T_DjiReturnCode DjiUpgradePlatformLinux_CleanUpgradeProgramFileStoreArea(void);

/* 创建升级文件 */
T_DjiReturnCode DjiUpgradePlatformLinux_CreateUpgradeProgramFile(const
T_DjiUpgradeFileInfo *fileInfo);

/* 写入升级文件 */
T_DjiReturnCode DjiUpgradePlatformLinux_WriteUpgradeProgramFile(uint32_t
offset, const uint8_t *data,uint16_t dataLen);
```

```c
/* 读取升级文件 */
T_DjiReturnCode DjiUpgradePlatformLinux_ReadUpgradeProgramFile(uint32_t
offset, uint16_t readDataLen, uint8_t *data,uint16_t *realLen);

/* 关闭升级文件 */
T_DjiReturnCode DjiUpgradePlatformLinux_CloseUpgradeProgramFile(void);

/* 替换升级文件 */

T_DjiReturnCode DjiUpgradePlatformLinux_ReplaceOldProgram(void);

/* 设置重启状态 */
T_DjiReturnCode DjiUpgradePlatformLinux_SetUpgradeRebootState(const
T_DjiUpgradeEndInfo *upgradeEndInfo);

/* 获取重启状态 */
T_DjiReturnCode DjiUpgradePlatformLinux_GetUpgradeRebootState(bool
*isUpgradeReboot, T_DjiUpgradeEndInfo *upgradeEndInfo);

/* 清除重启状态 */
T_DjiReturnCode DjiUpgradePlatformLinux_CleanUpgradeRebootState(void);

#ifdef __cplusplus
}
#endif

#endif // UPGRADE_PLATFORM_OPT_LINUX_H
/*********** (C) COPYRIGHT DJI Innovations *******END OF FILE******/
```

任务设计

本项目在已完成的"项目 9 任务 1 视频流文件传输与回放下载"的基础上，通过 PSDK 的本地升级功能，将负载设备的固件从版本 01.00.00.00 升级到版本 01.00.00.01。

任务的具体设计要求如下。

（1）在树莓派部署 FTP Server。

① 安装 VSFTP。

② 配置 VSFTP。

③ 设置 VSFTP 自启动。

（2）编译新固件。

① 在已完成的"项目 9 任务 1 视频流文件传输与回放下载"的基础上，加入本地升级功能。

② 修改固件版本为 01.00.00.01。

③ 修改新固件文件名。
(3) 设置负载设备自启动。
(4) 执行项目。
① 查看固件版本。
② 测试视频流文件传输与回放下载是否正常。

◆ 任务实施

限于篇幅，这里只给出任务实施步骤和关键代码。

1. 部署 FTP Server

本项目基于 FTP 进行本地升级，首先要在树莓派上部署 FTP Server。

(1) 安装 VSFTPD。

```
pi@raspberrypi:~ $ sudo apt-get install vsftpd
pi@raspberrypi:~ $
```

(2) 配置 VSFTPD。

```
pi@raspberrypi:~ $ sudo vim /etc/vsftpd.conf
................省略上面部分输出................
# Allow anonymous FTP? (Disabled by default).
anonymous_enable=NO
#
# Uncomment this to allow local users to log in.
local_enable=YES
#
# Uncomment this to enable any form of FTP write command.
write_enable=YES

................省略下面部分输出................
```

(3) 添加用户。

用户账号为 psdk_payload_ftp，密码为 DJi_#$31，该用户家目录为/upgrade。

```
pi@raspberrypi:~ $ sudo adduser psdk_payload_ftp --home /upgrade
Adding user `psdk_payload_ftp' ...
Adding new group `psdk_payload_ftp' (1001) ...
Adding new user `psdk_payload_ftp' (1001) with group `psdk_payload_ftp' ...
Creating home directory `/upgrade' ...
Copying files from `/etc/skel' ...
New password:DJi_#$31
Retype new password:DJi_#$31
passwd: password updated successfully
Changing the user information for psdk_payload_ftp
Enter the new value, or press ENTER for the default
        Full Name []:
        Room Number []:
        Work Phone []:
```

```
        Home Phone []:
        Other []:
Is the information correct? [Y/n] y
pi@raspberrypi:~ $

pi@raspberrypi:~ $ ls -l /
total 72
…
drwxr-xr-x 2 psdk_payload_ftp psdk_payload_ftp 4096 Feb 26 17:21 upgrade
…
pi@raspberrypi:~ $
```

(4)测试 FTP Server 登录。

在个人计算机上登录。192.168.0.199 是当前实验环境下的 IP 地址,大家需要根据自己的实验环境使用正确的 IP 地址。

```
C:\Users\Administrator>ftp 192.168.0.199
连接到 192.168.0.199。
220 (vsFTPd 3.0.3)
200 Always in UTF8 mode.
用户(192.168.0.199:(none)): psdk_payload_ftp
331 Please specify the password.
密码:DJi_#$31
230 Login successful.
ftp>
ftp> ls
200 PORT command successful. Consider using PASV.
150 Here comes the directory listing.
226 Directory send OK.
ftp> bye
221 Goodbye.

C:\Users\Administrator>
```

(5)设置 VSFTPD 开机启动。

```
# 在/etc/rc.local 文件中添加 /etc/init.d/vsftpd start
pi@raspberrypi:~ $ sudo vim /etc/rc.local
…
/etc/init.d/vsftpd start
…
```

2. 目录结构

本项目基于"项目 9 任务 1 视频流文件传输与回放下载",现在对负载设备固件版本进行升级,从版本 01.00.00.00 升级到版本 01.00.00.01。这里将"项目 9 任务 1 视频流文件传输与回放下载"的可运行的项目代码完整复制过来,并将项目命名为 pro_12_upgrade。

```
pi@raspberrypi:~ $ cp -r pro_9_video_playback pro_12_upgrade
pi@raspberrypi:~ $
```

增加本地升级功能后的文件目录结构如下。

```
pi@raspberrypi:~/pro_12_upgrade $ tree
.
├── 3rdparty
├── camera_emu                              # 相机功能
│   ├── dji_media_file_manage
│   │   ├── dji_media_file_core.c
│   │   ├── dji_media_file_core.h
│   │   ├── dji_media_file_jpg.c
│   │   ├── dji_media_file_jpg.h
│   │   ├── dji_media_file_mp4.c
│   │   └── dji_media_file_mp4.h
│   ├── media_file                          # 媒体文件
│   │   ├── out.h264                        # H.264 媒体流文件
│   │   ├── PSDK_0001_ORG.jpg               # 缩略图图片
│   │   ├── PSDK_0002_ORG.jpg               # 缩略图图片
│   │   ├── PSDK_0003_ORG.jpg               # 缩略图图片
│   │   ├── PSDK_0004_ORG.mp4               # MP4 回放下载文件
│   │   └── PSDK_0005.h264                  # H.264 媒体流文件
│   ├── payload_cam_emu_base.c              # 相机功能源文件
│   ├── payload_cam_emu_base.h              # 相机功能头文件
│   ├── payload_cam_emu_media.c             # 相机媒体功能源文件
│   └── payload_cam_emu_media.h             # 相机媒体功能头文件
├── CMakeLists.txt
├── dji_init.c
├── dji_init.h
├── gimbal_emu                              # 相机功能用到的云台功能
│   ├── payload_gimbal_emu.c                # 云台功能源文件
│   └── payload_gimbal_emu.h                # 云台功能头文件
├── hal
├── main.c
├── osal
├── psdk_lib
├── upgrade                                 # 本地需要用到目录和文件
│   ├── upgrade.c
│   ├── upgrade_common_file_transfer.c
│   ├── upgrade_common_file_transfer.h
│   ├── upgrade.h
│   ├── upgrade_platform_opt.c
│   └── upgrade_platform_opt.h
└── upgrade_platform_opt
    ├── upgrade_platform_opt_linux.c
    └── upgrade_platform_opt_linux.h
```

由于本项目基于项目 9 任务 1，因此所有关于项目 9 任务 1 的代码请参阅项目 9 任务 1，这里只关注项目 9 任务 1 的本地升级。

3. 修改 CMakeLists.txt

（1）设置本项目的工程名为 pro_12_upgrade。

```
# ------------------------设置工程名------------------------------
project(pro_12_upgrade C)
# ----------------------------------------------------------------
```

（2）添加本地升级需要用到的目录和源文件。

```
file(GLOB_RECURSE MODULE_UPGRADE_SRC upgrade/*.c)
include_directories(./upgrade)
file(GLOB_RECURSE MODULE_UPGRADE_PLATFORM_OPT_SRC upgrade_platform_opt/*.c)
include_directories(./upgrade_platform_opt)
# ----------------------------------------------------------------
```

（3）将本地升级需用到的目录和源文件添加到项目中。

```
add_executable(${PROJECT_NAME}
        main.c
        dji_init.c
        ${MODULE_OSAL_SRC}
        ${MODULE_HAL_SRC}
        ${MODULE_3RDPART_SRC}
        ${MODULE_MONITOR_SRC}
        ${MODULE_SAMPLE_SRC}
        ${MODULE_GIMBAL_SRC}
        ${MODULE_UPGRADE_SRC}
        ${MODULE_UPGRADE_PLATFORM_OPT_SRC}
        )
# ----------------------------------------------------------------
```

4. 修改 main.c

（1）添加头文件。

添加本地升级的两个头文件。

```c
#include <upgrade/upgrade.h>
#include <upgrade_platform_opt/upgrade_platform_opt_linux.h>
```

（2）添加本地升级服务初始化。

```c
int main(int argc, char **argv)
{
/* ------------以下是本地升级需要用到的功能----------------------*/
// 设置负载设备应用的别名，与项目的工程名一致
    returnCode = DjiCore_SetAlias("pro_12_upgrade");
    if (returnCode != DJI_ERROR_SYSTEM_MODULE_CODE_SUCCESS) {
        USER_LOG_ERROR("set alias error");
        return DJI_ERROR_SYSTEM_MODULE_CODE_SYSTEM_ERROR;
    }
// 设置升级操作
    T_DjiTestUpgradePlatformOpt linuxUpgradePlatformOpt = {
        .rebootSystem = DjiUpgradePlatformLinux_RebootSystem,
```

```
            .cleanUpgradeProgramFileStoreArea
=DjiUpgradePlatformLinux_CleanUpgradeProgramFileStoreArea,
            .createUpgradeProgramFile = DjiUpgradePlatformLinux_CreateUpgradeProgramFile,
            .writeUpgradeProgramFile =DjiUpgradePlatformLinux_WriteUpgradeProgramFile,
            .readUpgradeProgramFile =DjiUpgradePlatformLinux_ReadUpgradeProgramFile,
            .closeUpgradeProgramFile =DjiUpgradePlatformLinux_CloseUpgradeProgramFile,
            .replaceOldProgram = DjiUpgradePlatformLinux_ReplaceOldProgram,
            .setUpgradeRebootState DjiUpgradePlatformLinux_SetUpgradeRebootState,
            .getUpgradeRebootState = DjiUpgradePlatformLinux_GetUpgradeRebootState,
            .cleanUpgradeRebootState = DjiUpgradePlatformLinux_CleanUpgradeRebootState,
    };

    // 设置新固件版本为 01.00.00.01，默认版本为 01.00.00.00
        T_DjiTestUpgradeConfig testUpgradeConfig = {
                //.firmwareVersion = {1, 0, 0, 0},
            .firmwareVersion = {1, 0, 0, 1},
            .transferType = DJI_FIRMWARE_TRANSFER_TYPE_FTP,
            .needReplaceProgramBeforeReboot = true
    };
    // 在运行时查看版本，确认版本是否正确
        printf("Upgrade Firmware Version: V01.00.00.01");
        //printf("Initial Firmware Version: V01.00.00.00");
        if (DjiTest_UpgradeStartService(&linuxUpgradePlatformOpt, testUpgradeConfig) !=
            DJI_ERROR_SYSTEM_MODULE_CODE_SUCCESS) {
            USER_LOG_ERROR("psdk upgrade init error");
        }
    /* ------------添加本地升级功能结束----------------------*/
    }
```

5. 修改 upgrade_platform_opt_linux.h

修改 upgrade_platform_opt 目录下的头文件 upgrade_platform_opt_linux.h，设置老固件存放路径（"/home/pi/demo/bin/pro_12_upgrade"）。如果/home/pi/demo/bin 目录不存在，则创建该目录，并制定新固件，通过 FTP 将其下载至存放目录（"/upgrade/"），此目录是 FTP 用户 psdk_payload_ftp 的家目录，在创建用户时就被创建了。

```
#define DJI_TEST_UPGRADE_OLD_FIRMWARE_PATH      "/home/pi/demo/bin/pro_12_upgrade"
#define DJI_TEST_UPGRADE_FILE_DIR       "/upgrade/"
# --------------------------------------------------------------------
```

6. 操作过程

（1）先编译 01.00.00.00 版本。

在 main.c 中设置版本为{1,0,0,0}，代码如下。

```
        T_DjiTestUpgradeConfig testUpgradeConfig = {
            .firmwareVersion = {1, 0, 0, 0},
            .transferType = DJI_FIRMWARE_TRANSFER_TYPE_FTP,
            .needReplaceProgramBeforeReboot = true
```

```
};
printf("Initial Firmware Version: V01.00.00.00");
if (DjiTest_UpgradeStartService(&linuxUpgradePlatformOpt,
    testUpgradeConfig) != DJI_ERROR_SYSTEM_MODULE_CODE_SUCCESS) {
    USER_LOG_ERROR("psdk upgrade init error");
}
```

其他部分的代码不做任何修改。

（2）编译执行并查看版本。

编译执行后，初始版本如图 12-3 所示。

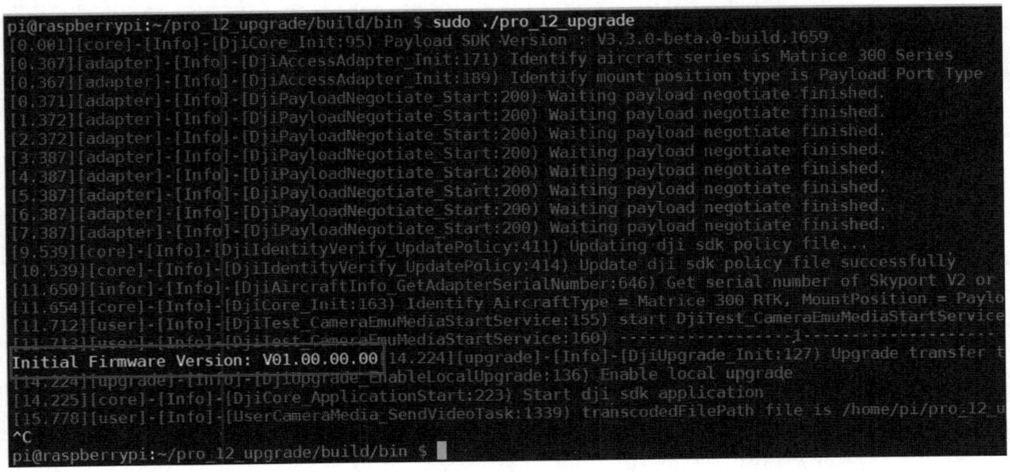

图 12-3　初始版本

（3）复制文件。

将项目的可视性文件复制到/home/pi/demo/bin 目录下，如果该目录不存在，则创建该目录。

```
pi@raspberrypi:~/pro_12_upgrade/build/bin $ mkdir -p /home/pi/demo/bin/
pi@raspberrypi:~/pro_12_upgrade/build/bin $ cp pro_12_upgrade /home/pi/demo/bin/

pi@raspberrypi:~/pro_12_upgrade/build/bin $ ls -l /home/pi/demo/bin/
total 2240
-rwxr-xr-x 1 pi   pi    757092 Feb 28 12:21 pro_12_upgrade
```

（4）编译新固件。

修改固件版本为{1,0,0,1}。

```
T_DjiTestUpgradeConfig testUpgradeConfig = {
    //.firmwareVersion = {1, 0, 0, 0},
    .firmwareVersion = {1, 0, 0, 1},
    .transferType = DJI_FIRMWARE_TRANSFER_TYPE_FTP,
    .needReplaceProgramBeforeReboot = true
};
printf("Upgrade Firmware Version: V01.00.00.01");
//printf("Initial Firmware Version: V01.00.00.00");
if (DjiTest_UpgradeStartService(&linuxUpgradePlatformOpt, testUpgradeConfig) !=
```

```
                        DJI_ERROR_SYSTEM_MODULE_CODE_SUCCESS) {
                        USER_LOG_ERROR("psdk upgrade init error");
                    }
```

（5）重新编译生成新固件。

```
pi@raspberrypi:~ /pro_12_upgrade/build $ sudo rm -rf *
pi@raspberrypi:~ /pro_12_upgrade/build $ cmake ..
pi@raspberrypi:~ /pro_12_upgrade/build $ make
```

（6）修改新固件文件名。

将新固件文件修改为 pro_12_upgrade_V01.00.00.01.bin。

```
pi@raspberrypi:~ /pro_12_upgrade/build $ cd bin
pi@raspberrypi:~ /pro_12_upgrade/build/bin $ cd ls -l
-rwxr-xr-x 1 pi pi 757092 Feb 28 12:25 pro_12_upgrade
pi@raspberrypi:~ /pro_12_upgrade/build/bin $ sudo pro_12_upgrade
pro_12_upgrade_V01.00.00.01.bin
```

（7）下载新固件文件。

将 pro_12_upgrade_V01.00.00.01.bin 下载到本地计算机中，并放置在 E:/upgrade 目录下。下载的新固件文件如图 12-4 所示。

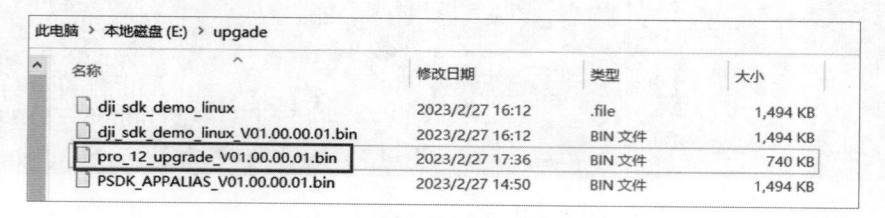

图 12-4　下载的新固件文件

7. 设置固件自启动

在/etc/rc.local 中添加如下内容。

```
chmod 666 /dev/ttyUSB0
chmod 755 /home/pi/demo/bin/pro_12_upgrade
sudo /home/pi/demo/bin/pro_12_upgrade &
```

ttyUSB0 为连接负载设备的串口设备，开发者可根据实际情况修改其名称。

```
pi@raspberrypi:~ $ ls /dev/ttyUSB0
/dev/ttyUSB0
```

8. 升级固件

（1）运行/home/pi/demo/bin 目录下的 pro_12_upgrade。

升级之前 pro_12_upgrade 的固件版本是 01.00.00.00，运行过程如图 12-3 所示。

（2）使用 DJI Assistant 2。

打开 DJI Assistant 2，选择编译生成的固件升级文件（pro_12_upgrade_V01.00.00.01.bin）进行固件升级。选择新版本，如图 12-5 所示。单击"升级"按钮，等待升级成功。升级成功后，如图 12-6 所示。

图 12-5　选择新版本

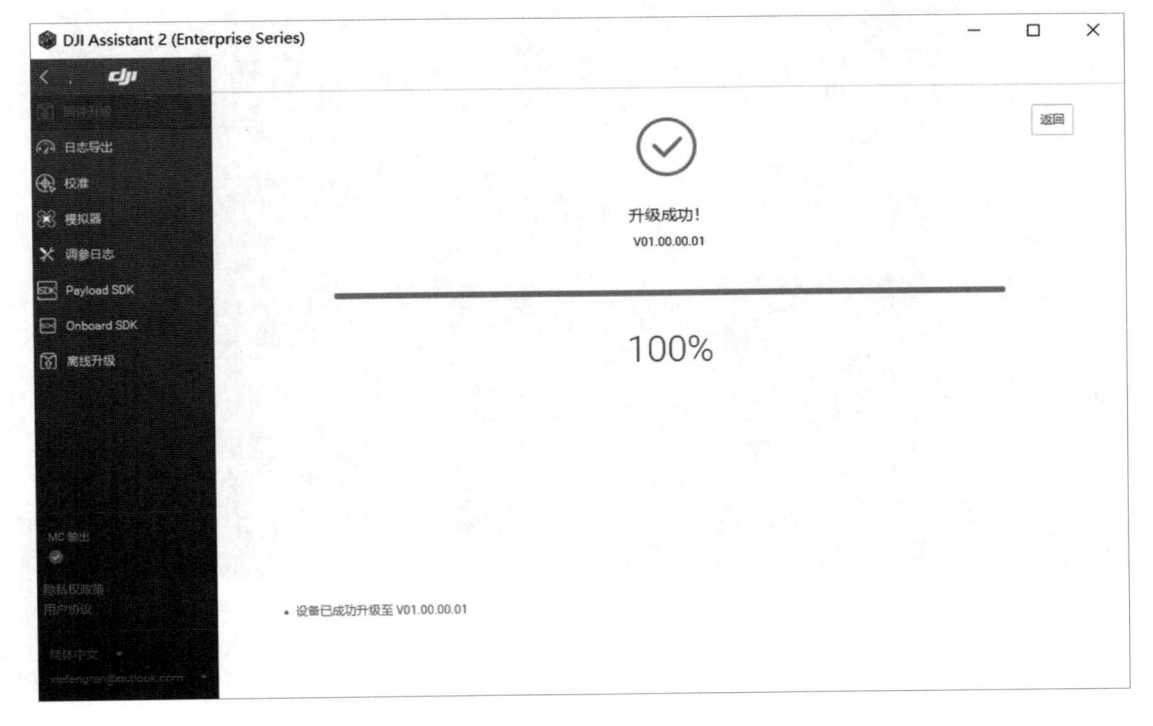

图 12-6　升级成功

(3) 测试。

树莓派重启后,在 Dji Polit 上可以看到项目 12 正常运行。通过下面的方法可以查看 pro_12_upgrade 的版本。

查看正在运行的 pro_12_upgrade 进程。

```
pi@raspberrypi:~ $ ps -aux |grep pro_12_upgrade
root   572  13352 4040 S14:56  0:00 sudo/home/pi/demo/bin/pro_12_upgrade
root   608 124552 2692 Sl14:56 0:11 /home/pi/demo/bin/pro_12_upgrade
pi    1437  7584  544 pts/0 S+ 14:58 0:00 grep --color=auto pro_12_upgrade
```

终止 pro_12_upgrade 进程。

```
pi@raspberrypi:~ $ sudo kill -9 572
pi@raspberrypi:~ $ sudo kill -9 608
pi@raspberrypi:~ $
```

进入 /home/pi/demo/bin 目录,运行 pro_12_upgrade。

```
pi@raspberrypi:~ $ cd demo/bin/
pi@raspberrypi:~/demo/bin $ ls -l
total 748
drwxr-xr-x 2 root root   4096 Feb 28 14:47 Logs
-rwxr-xr-x 1 root root 757092 Feb 28 14:49 pro_12_upgrade
-rw-r--r-- 1 root root      4 Feb 28 14:49 reboot_state
```

观察到版本号变为 01.00.00.01,说明升级成功。查看升级后的版本,如图 12-7 所示。

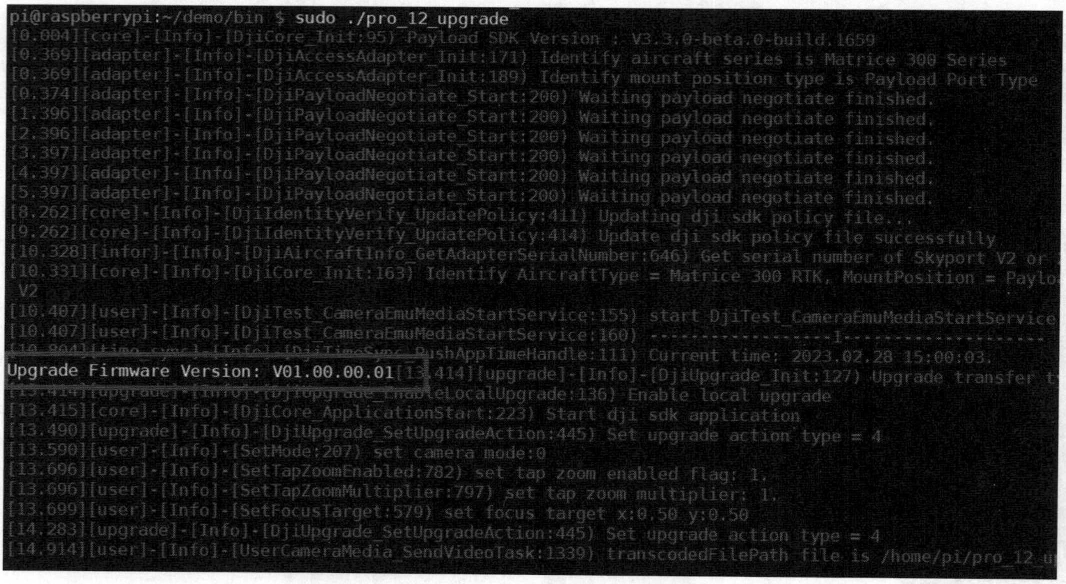

图 12-7 查看升级后的版本

任务评价

任务过程评价表

任务实施人姓名_____ 学号_____ 时间_____

	评价项目及标准	分值	小组评议	教师评议
技术能力	1. 对基本概念的熟悉程度	10		
	2. 掌握在树莓派上部署 FTP Server 的方法	10		
	3. 掌握配置和设置 FTP Server 自启动的方法	10		
	4. 掌握控制负载设备版本的方法	10		
	5. 掌握使用 DJI Assistant 2 升级固件的方法	10		
	6. 掌握本地升级的测试方法	10		
执行能力	1. 出勤情况	5		
	2. 遵守纪律情况	5		
	3. 是否主动参与，有无提问记录	5		
	4. 有无职业意识	5		
社会能力	1. 能否有效沟通	5		
	2. 是否使用基本的文明礼貌用语	5		
	3. 能否与组员主动交流、积极合作	5		
	4. 能否自我学习、自我管理	5		
总分		100		
评定等级：				
评价意见		学习意见		

评定等级具体如下。

A：优，得分>90。B：好，90≥得分>80。C：一般，80≥得分>60。D：有待提高，得分≤60。

小结

本项目首先简要介绍了 PSDK 本地升级的基本概念；其次介绍了本地升级的传输协议、固件版本控制，以及相关的接口说明；最后介绍了如何结合 DJI Assistant 2 对固件进行升级。

项目13 相机实现与云台实现

任务要求

本项目使用 PSDK 提供的相机功能，并通过负载设备（树莓派）上的摄像头实现拍照、录像等相机基础功能。

知识导入

13.1 概述

为满足开发者对相机类负载设备的控制需求，PSDK 提供了控制相机执行拍照、录像、变焦及对焦等功能的接口。开发者需要先实现相机拍照、录像及测光等功能，再通过注册 PSDK 相机类的接口，开发出功能完善的相机类负载设备；通过使用 DJI Pilot 或基于 MSDK 开发的移动端 App，用户能够控制基于 PSDK 开发的相机类负载设备执行指定的动作，并获取负载设备中的信息和资源。

13.2 相机功能

相机功能可分为基础功能和高级功能。
（1）基础功能：设置相机模式、拍照、录像、获取相机状态、SD 卡管理。
（2）高级功能：变焦、测光、对焦、视频流传输、回放下载、媒体库管理。

注意：本项目基于树莓派上的摄像头，只能实现相机的基础功能，无法完成相机的高级功能。下面介绍的内容也是基于相机的基础功能的，不涉及相机的高级功能。

13.2.1 基本概念

1. 相机模式

在使用相机类功能前，需要先设置相机类负载设备的模式，不同的模式指定了相机类负载设备在执行某个任务时的工作逻辑。

（1）拍照：在该模式下，用户能够触发相机类负载设备拍摄照片。

（2）录像：在该模式下，用户能够触发相机类负载设备录制影像。

（3）视频回放：在该模式下，用户可以在移动端 App 上播放或下载负载设备上的媒体文件。

注意：相机只能在一种模式下执行相应的操作，如在录像模式下仅能录像，无法拍照。

2. 拍照模式

使用 PSDK 开发的相机类负载设备支持以下拍照模式。

（1）单拍：下发拍照命令后，相机将拍摄单张照片。

（2）连拍：下发拍照命令后，相机将连续拍摄指定数量的照片；当前支持 2/3/5/7/10/14 张连拍。

（3）定时拍照：下发拍照命令后，相机将按照指定的时间间隔拍摄指定数量的照片。

- 当前支持 2/3/5/7/10 秒时间间隔。
- 当前最多支持指定拍摄 254 张照片，当拍摄张数达到 255 时，相机将不间断地拍摄照片。

3. 媒体文件

基于 PSDK 开发的相机类负载设备能够根据用户的指令，执行删除或下载文件等操作。支持用户使用 DJI Pilot 或基于 MSDK 开发的移动端 App 预览负载设备中的媒体文件。

1）静态预览

静态预览是指预览单个文件或文件列表。

（1）缩略图：预览文件列表。

- 图像：负载设备按照文件的原始比例生成缩略图，请将预览图的宽度设置为 100 像素。
- 视频：截取视频某一帧的画面。

（2）截屏图：预览单个文件。

- 图像：按照原始比例，建议将图像设置成宽为 600 像素的预览图。
- 视频：截取视频某一帧的画面。

（3）原始文件，如需获取相机类负载设备中原始的媒体文件，则请使用下载功能获取指定的媒体文件。

2）动态预览（视频预览）

动态预览是指播放、暂停、停止、跳转（快进、快退和进度拖动）。支持动态预览的文件格式有 MP4、JPG、DNG 和 MOV，编码格式请参见视频标准。

13.2.2 注册基础功能

开发者根据选用的开发平台及行业应用实际的使用需求，按照 PSDK 中的结构体 T_DjiCameraCommonHandler 构造并实现相机类负载设备设置相机模式、拍照和录像等功能的函数，将相机功能的函数注册到 PSDK 中的指定接口后，用户通过使用 DJI Pilot 或基于 MSDK 开发的移动端 App 能够控制基于 PSDK 开发的相机类负载设备执行相应的动作。

1. 功能结构体

PSDK 中的结构体 T_DjiCameraCommonHandler 用于构造并实现相机类负载设备设置相机模式、拍照和录像等功能的函数。为了控制相机类负载设备执行基础动作，根据本结构体中的函数构造了相机类负载设备执行基础功能的回调函数，其中包括：

- 获取相机类负载设备当前状态的回调函数。
- 设置相机类负载设备工作模式的回调函数。
- 获取相机类负载设备当前工作模式的回调函数。
- 控制相机类负载设备开始录像的回调函数。
- 控制相机类负载设备停止录像的回调函数。
- 控制相机类负载设备开始拍摄照片的回调函数。
- 控制相机类负载设备停止拍摄照片的回调函数。
- 设置相机类负载设备拍照模式的回调函数。
- 获取相机类负载设备当前拍照模式的回调函数。
- 设置相机类负载设备连拍模式下连拍张数的回调函数。
- 获取相机类负载设备当前连拍张数的回调函数。
- 设置相机类负载设备定时拍照模式下拍照间隔的回调函数。
- 获取相机类负载设备定时拍照模式下拍照间隔的回调函数。
- 获取相机类负载设备中当前 SD 卡状态的回调函数。
- 控制相机类负载设备格式化 SD 卡的回调函数。

结构体定义如下：

```
typedef struct {
  // 获取相机类负载设备当前状态的回调函数
  T_DjiReturnCode (*GetSystemState)(T_DjiCameraSystemState *systemState);

  // 设置相机类负载设备工作模式的回调函数
  T_DjiReturnCode (*SetMode)(E_DjiCameraMode mode);

  // 获取相机类负载设备当前工作模式的回调函数
  T_DjiReturnCode (*GetMode)(E_DjiCameraMode *mode);

  // 控制相机类负载设备开始录像的回调函数
  T_DjiReturnCode (*StartRecordVideo)(void);

  // 控制相机类负载设备停止录像的回调函数
```

```
    T_DjiReturnCode (*StopRecordVideo)(void);

    // 控制相机类负载设备开始拍摄照片的回调函数
    T_DjiReturnCode (*StartShootPhoto)(void);

    // 控制相机类负载设备停止拍摄照片的回调函数
    T_DjiReturnCode (*StopShootPhoto)(void);

    // 设置相机类负载设备拍照模式的回调函数
    T_DjiReturnCode (*SetShootPhotoMode)(E_DjiCameraShootPhotoMode mode);

    // 获取相机类负载设备当前拍照模式的回调函数
    T_DjiReturnCode (*GetShootPhotoMode)(E_DjiCameraShootPhotoMode *mode);

    // 设置相机类负载设备连拍模式下连拍张数的回调函数
    T_DjiReturnCode (*SetPhotoBurstCount)(E_DjiCameraBurstCount burstCount);

    // 获取相机类负载设备当前连拍张数的回调函数
    T_DjiReturnCode (*GetPhotoBurstCount)(E_DjiCameraBurstCount *burstCount);

    // 设置相机类负载设备定时拍照模式下拍照间隔的回调函数
    T_DjiReturnCode (*SetPhotoTimeIntervalSettings)
(T_DjiCameraPhotoTimeIntervalSettings settings);

    // 获取相机类负载设备定时拍照模式下拍照间隔的回调函数
    T_DjiReturnCode (*GetPhotoTimeIntervalSettings)
(T_DjiCameraPhotoTimeIntervalSettings *settings);

    // 获取相机类负载设备中当前 SD 卡状态的回调函数
    T_DjiReturnCode (*GetSDCardState)(T_DjiCameraSDCardState *sdCardState);

    // 控制相机类负载设备格式化 SD 卡的回调函数
    T_DjiReturnCode (*FormatSDCard)(void);

} T_DjiCameraCommonHandler;
```

为了防止以上函数阻塞 PSDK 的主线程，从而导致程序响应缓慢、相机类负载设备断连及死循环等问题，建议不要以阻塞的方式在回调函数中执行以上函数。在使用控制相机类负载设备执行基础动作的功能时，请根据结构体中的函数原型开发相机类负载设备的基础功能。

2. 注册相机类基础功能

开发者在实现相机类负载设备设置相机模式、拍照和录像等功能后，需要通过 DjiPayloadCamera_RegCommonHandler 注册相机类基础功能。

```
    static T_DjiCameraCommonHandler s_commonHandler;

    // 获取负载设备系统当前的状态
```

```c
s_commonHandler.GetSystemState = GetSystemState;

// 实现设置相机类负载设备模式的功能
s_commonHandler.SetMode = SetMode;
s_commonHandler.GetMode = DjiTest_CameraGetMode;

// 实现开始或停止录像的功能
s_commonHandler.StartRecordVideo = StartRecordVideo;
s_commonHandler.StopRecordVideo = StopRecordVideo;

// 实现开始或停止拍照的功能
s_commonHandler.StartShootPhoto = StartShootPhoto;
s_commonHandler.StopShootPhoto = StopShootPhoto;

// 实现设置相机类负载设备的拍照功能
s_commonHandler.SetShootPhotoMode = SetShootPhotoMode;
s_commonHandler.GetShootPhotoMode = GetShootPhotoMode;
s_commonHandler.SetPhotoBurstCount = SetPhotoBurstCount;
s_commonHandler.GetPhotoBurstCount = GetPhotoBurstCount;
s_commonHandler.SetPhotoTimeIntervalSettings = SetPhotoTimeIntervalSettings;
s_commonHandler.GetPhotoTimeIntervalSettings = GetPhotoTimeIntervalSettings;

// 实现 SD 卡管理功能
s_commonHandler.GetSDCardState = GetSDCardState;
s_commonHandler.FormatSDCard = FormatSDCard;

returnCode = DjiPayloadCamera_RegCommonHandler(&s_commonHandler);
if (returnCode != DJI_ERROR_SYSTEM_MODULE_CODE_SUCCESS) {
    USER_LOG_ERROR("camera register common handler error:0x%08llX", returnCode);
}
```

13.2.3 拍照功能

在使用拍照功能前，用户需要在 DJI Pilot 或基于 MSDK 开发的移动端 App 上将相机类负载设备的工作模式设置为拍照模式。基于 PSDK 开发的负载设备在拍照时，会向 DJI Pilot 或基于 MSDK 开发的移动端 App 返回拍照状态（用于触发移动端 App 拍照声音等功能）。

基于 PSDK 开发的负载设备控制程序调用 SetShootPhotoMode 和 GetShootPhotoMode 函数能够设置并获取相机类负载设备的模式。同时，用户使用 DJI Pilot 或基于 MSDK 开发的移动端 App 能够设置并获取相机类负载设备的拍照模式。

```c
staticT_DjiReturnCode SetShootPhotoMode(E_DjiCameraShootPhotoMode mode)
{
    T_DjiReturnCode returnCode;
    T_DjiOsalHandler *osalHandler = DjiPlatform_GetOsalHandler();

    returnCode = osalHandler->MutexLock(s_commonMutex);
```

```
    if (returnCode != DJI_ERROR_SYSTEM_MODULE_CODE_SUCCESS) {
        USER_LOG_ERROR("lock mutex error: 0x%08llX.", returnCode);
        return returnCode;
    }

    s_cameraShootPhotoMode = mode;
    USER_LOG_INFO("set shoot photo mode:%d", mode);

    returnCode = osalHandler->MutexUnlock(s_commonMutex);
    if (returnCode != DJI_ERROR_SYSTEM_MODULE_CODE_SUCCESS) {
        USER_LOG_ERROR("unlock mutex error: 0x%08llX.", returnCode);
        return returnCode;
    }

    return DJI_ERROR_SYSTEM_MODULE_CODE_SUCCESS;
}

Static T_DjiReturnCode GetShootPhotoMode(E_DjiCameraShootPhotoMode *mode)
{
    T_DjiReturnCode returnCode;
    T_DjiOsalHandler *osalHandler = DjiPlatform_GetOsalHandler();

    returnCode = osalHandler->MutexLock(s_commonMutex);
    if (returnCode != DJI_ERROR_SYSTEM_MODULE_CODE_SUCCESS) {
        USER_LOG_ERROR("lock mutex error: 0x%08llX.", returnCode);
        return returnCode;
    }

    *mode = s_cameraShootPhotoMode;

    returnCode = osalHandler->MutexUnlock(s_commonMutex);
    if (returnCode != DJI_ERROR_SYSTEM_MODULE_CODE_SUCCESS) {
        USER_LOG_ERROR("unlock mutex error: 0x%08llX.", returnCode);\
        return returnCode;
    }

    return DJI_ERROR_SYSTEM_MODULE_CODE_SUCCESS;
}
```

13.2.4 存储照片

相机类负载设备在执行完拍照动作后，基于 PSDK 开发的相机类负载设备会将相机拍摄的照片存储在 SD 卡中。

```
// 存储单拍模式下相机类负载设备拍摄的照片
```

```
        if (s_cameraShootPhotoMode == DJI_CAMERA_SHOOT_PHOTO_MODE_SINGLE) {
            s_cameraSDCardState.remainSpaceInMB =
            s_cameraSDCardState.remainSpaceInMB - SDCARD_PER_PHOTO_SPACE_IN_MB;
                                    s_cameraState.isStoring = false;
            s_cameraState.shootingState = DJI_CAMERA_SHOOTING_PHOTO_IDLE;
        }
        // 存储连拍模式下相机类负载设备拍摄的照片
        else if (s_cameraShootPhotoMode == DJI_CAMERA_SHOOT_PHOTO_MODE_BURST) {
            s_cameraSDCardState.remainSpaceInMB =
            s_cameraSDCardState.remainSpaceInMB - SDCARD_PER_PHOTO_SPACE_IN_MB *
s_cameraBurstCount;
            s_cameraState.isStoring = false;
            s_cameraState.shootingState = DJI_CAMERA_SHOOTING_PHOTO_IDLE;
        }
        // 存储定时拍照模式下相机类负载设备拍摄的照片
        else if (s_cameraShootPhotoMode == DJI_CAMERA_SHOOT_PHOTO_MODE_INTERVAL) {
            if (isStartIntervalPhotoAction == true) {
                s_cameraState.isStoring = false;
                s_cameraState.shootingState = DJI_CAMERA_SHOOTING_PHOTO_IDLE;
                s_cameraSDCardState.remainSpaceInMB =
                s_cameraSDCardState.remainSpaceInMB - SDCARD_PER_PHOTO_SPACE_IN_MB;
            }
        }
```

13.2.5 录像功能

由于相机类负载设备在录像的过程中无法拍照和测光，因此开发者可根据用户的使用需要，设置相机类负载设备录像时 ISO、曝光、对焦等参数的默认值。在使用相机类负载设备的录像功能前，用户需要在 DJI Pilot 或基于 MSDK 开发的移动端 App 上将相机类负载设备的模式设置为录像模式。

基于 PSDK 开发的负载设备控制程序调用 StartRecordVideo 函数和 StopRecordVideo 函数，能够控制相机类负载设备录像。同时，用户使用 DJI Pilot 或基于 MSDK 开发的移动端 App 能够控制相机类负载设备录像。

```
static T_DjiReturnCode StartRecordVideo(void)
{
    T_DjiReturnCode djiStat;
    T_DjiReturnCode returnCode = DJI_ERROR_SYSTEM_MODULE_CODE_SUCCESS;
    T_DjiOsalHandler *osalHandler = DjiPlatform_GetOsalHandler();

    djiStat = osalHandler->MutexLock(s_commonMutex);
    if (djiStat != DJI_ERROR_SYSTEM_MODULE_CODE_SUCCESS) {
        USER_LOG_ERROR("lock mutex error: 0x%08llX.", djiStat);
        return djiStat;
    }
```

```c
    if (s_cameraState.isRecording != false) {
        USER_LOG_ERROR("camera is already in recording state");
        returnCode = DJI_ERROR_SYSTEM_MODULE_CODE_NONSUPPORT_IN_CURRENT_STATE;
        goto out;
    }

    s_cameraState.isRecording = true;
    USER_LOG_INFO("start record video");

out:
    djiStat = osalHandler->MutexUnlock(s_commonMutex);
    if (djiStat != DJI_ERROR_SYSTEM_MODULE_CODE_SUCCESS) {
        USER_LOG_ERROR("unlock mutex error: 0x%08llX.", djiStat);
        return djiStat;
    }

    return returnCode;
}

static T_DjiReturnCode StopRecordVideo(void)
{
    T_DjiReturnCode djiStat;
    T_DjiReturnCode returnCode = DJI_ERROR_SYSTEM_MODULE_CODE_SUCCESS;
    T_DjiOsalHandler *osalHandler = DjiPlatform_GetOsalHandler();

    djiStat = osalHandler->MutexLock(s_commonMutex);
    if (djiStat != DJI_ERROR_SYSTEM_MODULE_CODE_SUCCESS) {
        USER_LOG_ERROR("lock mutex error: 0x%08llX.", djiStat);
        return djiStat;
    }

    if (s_cameraState.isRecording != true) {
        USER_LOG_ERROR("camera is not in recording state");
        returnCode = DJI_ERROR_SYSTEM_MODULE_CODE_NONSUPPORT_IN_CURRENT_STATE;
        goto out;
    }

    s_cameraState.isRecording = false;
    s_cameraState.currentVideoRecordingTimeInSeconds = 0;
    USER_LOG_INFO("stop record video");

out:
    djiStat = osalHandler->MutexUnlock(s_commonMutex);
    if (djiStat != DJI_ERROR_SYSTEM_MODULE_CODE_SUCCESS) {
        USER_LOG_ERROR("unlock mutex error: 0x%08llX.", djiStat);
        return djiStat;
    }
```

```
    return returnCode;
}
```

基于 PSDK 开发的相机类负载设备控制程序，默认以 10Hz 的频率更新相机的状态。相机开始录像后，DJI Pilot 或基于 MSDK 开发的移动端 App 会显示当前正在录像的时间，当相机停止录像时，该时间将归 0。

13.3 云台功能

本项目不包含云台功能的实现，以下内容仅介绍云台功能的基本概念。

要想使用 PSDK 的云台功能，开发者需要先设计负载设备的云台并开发出控制云台的程序，将云台的控制函数注册到 PSDK 指定的接口后，再通过 DJI Pilot、基于 MSDK 开发的移动端 App 及遥控器，即可控制基于 PSDK 开发的具有云台功能的负载设备，同时获取负载设备的相关信息，如姿态等。

基于 PSDK 开发的云台类负载设备需要按照指定的要求上报云台的状态、当前姿态和校准状态等信息，方便用户在移动端 App 或机载计算机上根据云台的状态实现精准控制。

1. 云台关节与云台关节角

云台关节是指云台上带动负载设备转动的结构件，即云台电机；云台关节角是指云台电机转动的角度。关于云台关节的详细描述请参阅大疆官网。

2. 云台姿态与云台姿态角

云台姿态是指根据用户的控制指令，云台能够调整姿态；云台姿态角是指使用大地坐标系描述云台上负载设备的角度，该角度也被称为欧拉角。关于云台姿态与云台姿态角的详细描述请参阅大疆官网。

3. 云台模式

云台模式决定了云台跟随无人机运动时的转动方式。
- 自由模式：当无人机的姿态发生改变时，云台将不会转动。
- FPV 模式：当无人机的姿态发生改变时，云台将会转动航向轴与横滚轴，以确保负载设备当前的视场角不会发生改变。
- YAW 跟随模式：在该模式下，云台的航向轴将会跟随无人机的航向轴转动。

说明：在以上 3 种模式下，无人机系统中的其他模块（航线飞行时的云台控制指令）、遥控器和移动端 App 都能够控制云台转动。

4. 云台控制

1）控制模式

云台转动的控制模式分为以下 3 种。
- 相对角度控制：基于 PSDK 开发的云台根据用户指定的角度，在规定的时间内，转动指定的角度。
- 绝对角度控制：基于 PSDK 开发的云台根据用户的指令，在规定的时间内，从当前的位置转动到指定的位置。
- 速度控制：用户可控制基于 PSDK 开发的云台的转动速度。

说明：在上述两种角度控制模式下，云台转动的时间受云台最大旋转速度和最大加速度的限制，实际的转动角度受云台限位角的限制。在速度控制模式下，云台将根据用户指定的速度转动 0.5s，当云台转动到限位角时，将停止转动。

2）控制规则

- 优先级低的控制动作在优先级高的控制动作结束后才能控制云台。
- 优先级高的控制动作可抢占优先级低的控制动作的控制权。
- 同等优先级的控制场景按照开始控制的时间先后顺序抢占控制权，开始控制时间较晚的控制场景不能夺取控制权。

3）权限释放

- 控制模块完成对云台的控制后会释放控制权。
- 若控制模块完成对云台的控制后未释放控制权，则基于 PSDK 开发的云台将在云台转动结束后的指定时间内自动释放控制权。

5. 云台复位

基于 PSDK 开发的云台类负载设备支持用户通过 DJI Pilot 或基于 MSDK 开发的移动端 App 复位云台，将云台的姿态复位为初始状态。

- 航向轴复位：将云台航向轴的角度复位为无人机航向轴角度与云台航向轴微调角度的和。
- 俯仰轴与航向轴复位：将云台俯仰轴的角度复位为微调的角度，将云台航向轴的角度复位为无人机航向轴角度与云台航向轴微调角度的和。
- 重置云台的偏航轴和俯仰轴：将云台偏航轴的角度重置为无人机偏航轴和云台微调角度的和。重置云台俯仰轴为-90°与云台微调角度的和（云台下置）、90°与云台微调角度的和（云台上置）。

13.4 代码模板

1. 代码结构

下面是实现基础相机功能的代码框架，用户只需要编写用于实现基础相机功能的 payload_cam_emu_base.c 和 payload_cam_emu_base.h 即可。

```
.
├── 3rdparty
├── camera_emu
```

```
        |       ├── payload_cam_emu_base.c           # 相机基础功能源文件
        |       └── payload_cam_emu_base.h           # 头文件
        ├── gimbal_emu
        ├── hal
        ├── osal
        ├── psdk_lib
        └── raspi_camera
            ├── include
            |   └── RaspiCam.h
            └── lib
                └── libRaspiCamera.a
```

2. 头文件模板

本项目的头文件名为 payload_cam_emu_base.h。该文件定义的内容可满足正常的应用开发需求，如果需要实现特殊的功能，则可以按需要进行修改。

这里不再展示头文件代码。

3. 源文件模板

本项目的源文件名为 payload_cam_emu_base.c，以下是 payload_cam_emu_base.c 的模板文件，按照提示即可完成代码的编写。

由于源文件的代码行数太多，这里仅给出在 payload_cam_emu_base.c 文件中需要完成的功能。

（1）在服务入口函数中完成以下功能。

```c
/* 在相机功能服务入口函数 DjiTest_CameraEmuBaseStartService 中按照提示完成以下功能--------*/
/* step 1 调用 DjiPayloadCamera_Init 函数初始化相机类负载设备*/
/* step 2 在结构体 T_DjiCameraCommonHandler 对象 s_commonHandler 中添加拍照和录像功能*/
/* step 3 调用 DjiPayloadCamera_RegCommonHandler 函数注册相机基础功能*/

T_DjiReturnCode DjiTest_CameraEmuBaseStartService(void)
{
    T_DjiReturnCode returnCode;
    char ipAddr[DJI_IP_ADDR_STR_SIZE_MAX] = {0};
    uint16_t port = 0;
    T_DjiOsalHandler *osalHandler = DjiPlatform_GetOsalHandler();
    T_DjiAircraftInfoBaseInfo aircraftInfoBaseInfo = {0};

    returnCode = osalHandler->MutexCreate(&s_commonMutex);
    if (returnCode != DJI_ERROR_SYSTEM_MODULE_CODE_SUCCESS) {
        USER_LOG_ERROR("create mutex used to lock tap zoom arguments error: 0x%08llX", returnCode);
        return returnCode;
    }
```

```c
    returnCode = osalHandler->MutexCreate(&s_zoomMutex);
    if (returnCode != DJI_ERROR_SYSTEM_MODULE_CODE_SUCCESS) {
        USER_LOG_ERROR("create mutex used to lock tap zoom arguments error: 0x%08llX", returnCode);
        return returnCode;
    }

    returnCode = osalHandler->MutexCreate(&s_tapZoomMutex);
    if (returnCode != DJI_ERROR_SYSTEM_MODULE_CODE_SUCCESS) {
        USER_LOG_ERROR("create mutex used to lock tap zoom arguments error: 0x%08llX", returnCode);
        return returnCode;
    }
    /* step 1 调用DjiPayloadCamera_Init函数初始化相机类负载设备 -----*/

    /* ----------------------------------------------------------*/

    s_cameraSDCardState.isInserted = true;
    s_cameraSDCardState.totalSpaceInMB = SDCARD_TOTAL_SPACE_IN_MB;
    s_cameraSDCardState.remainSpaceInMB = SDCARD_TOTAL_SPACE_IN_MB;
    s_cameraSDCardState.availableCaptureCount = SDCARD_TOTAL_SPACE_IN_MB / SDCARD_PER_PHOTO_SPACE_IN_MB;
    s_cameraSDCardState.availableRecordingTimeInSeconds =
        SDCARD_TOTAL_SPACE_IN_MB / SDCARD_PER_SECONDS_RECORD_SPACE_IN_MB;

    /* step 2 在结构体T_DjiCameraCommonHandler对象s_commonHandler中添加拍照和录像
功能*/
    s_commonHandler.StartShootPhoto =    ;
    s_commonHandler.StopShootPhoto =    ;
    s_commonHandler.StartRecordVideo =    ;
    s_commonHandler.StopRecordVideo =    ;
    /* ----------------------------------------------------------*/

    s_commonHandler.GetSystemState = GetSystemState;
    s_commonHandler.SetMode = SetMode;
    s_commonHandler.GetMode = DjiTest_CameraGetMode;
    s_commonHandler.SetShootPhotoMode = SetShootPhotoMode;
    s_commonHandler.GetShootPhotoMode = GetShootPhotoMode;
    s_commonHandler.SetPhotoBurstCount = SetPhotoBurstCount;
    s_commonHandler.GetPhotoBurstCount = GetPhotoBurstCount;
    s_commonHandler.SetPhotoTimeIntervalSettings = SetPhotoTimeIntervalSettings;
    s_commonHandler.GetPhotoTimeIntervalSettings = GetPhotoTimeIntervalSettings;
    s_commonHandler.GetSDCardState = GetSDCardState;
    s_commonHandler.FormatSDCard = FormatSDCard;
```

```
        /* step 3 调用 DjiPayloadCamera_RegCommonHandler 函数注册相机基础功能-*/



        /* -------------------------------------------------------------*/
}
```

(2) 在拍照函数 StartShootPhoto 中添加拍照功能。
```
static T_DjiReturnCode StartShootPhoto(void)
{
    T_DjiReturnCode returnCode;
    T_DjiOsalHandler *osalHandler = DjiPlatform_GetOsalHandler();
    char cmdStr[128];

    returnCode = osalHandler->MutexLock(s_commonMutex);
    if (returnCode != DJI_ERROR_SYSTEM_MODULE_CODE_SUCCESS) {
        USER_LOG_ERROR("lock mutex error: 0x%08llX.", returnCode);
        return returnCode;
    }

    USER_LOG_INFO("start shoot photo");
    s_cameraState.isStoring = true;

/* ---在此处添加拍照功能----------------------------------------*/


    /* -------------------------------------------------------------*/
    } else if (s_cameraShootPhotoMode == DJI_CAMERA_SHOOT_PHOTO_MODE_BURST) {
        s_cameraState.shootingState = DJI_CAMERA_SHOOTING_BURST_PHOTO;
    } else if (s_cameraShootPhotoMode == DJI_CAMERA_SHOOT_PHOTO_MODE_INTERVAL) {
        s_cameraState.shootingState = DJI_CAMERA_SHOOTING_INTERVAL_PHOTO;
        s_cameraState.isShootingIntervalStart = true;
        s_cameraState.currentPhotoShootingIntervalTimeInSeconds =
s_cameraPhotoTimeIntervalSettings.timeIntervalSeconds;
    }

    returnCode = osalHandler->MutexUnlock(s_commonMutex);
    if (returnCode != DJI_ERROR_SYSTEM_MODULE_CODE_SUCCESS) {
        USER_LOG_ERROR("unlock mutex error: 0x%08llX.", returnCode);
        return returnCode;
    }

    return DJI_ERROR_SYSTEM_MODULE_CODE_SUCCESS;
}
```

(3) 在录像函数 StartRecordVideo 中添加录像功能。
```
static T_DjiReturnCode StartRecordVideo(void)
{
```

```c
    T_DjiReturnCode djiStat;
    T_DjiReturnCode returnCode = DJI_ERROR_SYSTEM_MODULE_CODE_SUCCESS;
    T_DjiOsalHandler *osalHandler = DjiPlatform_GetOsalHandler();
    char cmdStr[128];

    djiStat = osalHandler->MutexLock(s_commonMutex);
    if (djiStat != DJI_ERROR_SYSTEM_MODULE_CODE_SUCCESS) {
        USER_LOG_ERROR("lock mutex error: 0x%08llX.", djiStat);
        return djiStat;
    }

    if (s_cameraState.isRecording != false) {
        USER_LOG_ERROR("camera is already in recording state");
        returnCode = DJI_ERROR_SYSTEM_MODULE_CODE_NONSUPPORT_IN_CURRENT_STATE;
        goto out;
    }

    s_cameraState.isRecording = true;
    USER_LOG_INFO("start record video");

/* ---添加录像功能------------------------------------------------*/



    /*-----------------------------------------------------------*/

out:
    djiStat = osalHandler->MutexUnlock(s_commonMutex);
    if (djiStat != DJI_ERROR_SYSTEM_MODULE_CODE_SUCCESS) {
        USER_LOG_ERROR("unlock mutex error: 0x%08llX.", djiStat);
        return djiStat;
    }
    return returnCode;
}
```

任务设计

本项目使用 PSDK 提供的基础相机功能在负载设备（树莓派+摄像头）上实现拍照和录像功能。

任务的具体设计要求如下。

（1）从 pro_13_camera_source_template 项目模板文件中复制一个新项目。项目的工程名为 pro_13_camera。

（2）本项目将实现拍照和录像功能，项目目录下的 camera_emu 目录用于编写代码，以实现拍照和录像功能 API 接口。

（3）以自动的方式设置网络参数。

（4）在 camera_emu 目录下编写代码，实现拍照和录像功能。
① 初始化相机功能。
② 注册相机功能函数。
③ 实现拍照功能。
④ 实现录像功能。
（5）使用 cmake 编译项目。
（6）执行项目。
① 测试拍照功能。
② 测试录像功能。

任务实施

限于篇幅，这里只给出任务实施步骤和关键代码。

1. 模板文件

连接树莓派开发板，默认路径为当前登录用户的家目录。查看当前的家目录，该目录下有一个名为 pro_13_camera_source_template 的目录，此目录下保存了本项目的所有模板文件。目录结构如下。

```
pi@raspberrypi:~ $ cd pro_13_camera_source_template/
pi@raspberrypi:~/ pro_13_camera_source_template $ tree
.
├── 3rdparty
├── camera_emu
│   ├── payload_cam_emu_base.c         # 相机基础功能源文件
│   └── payload_cam_emu_base.h         # 头文件
├── gimbal_emu
├── hal
├── osal
├── psdk_lib
└── raspi_camera
    ├── include
    │   └── RaspiCam.h
    └── lib
        └── libRaspiCamera.a
```

2. 复制模板目录

将 pro_13_camera_source_template 目录下的所有文件和目录复制到 pro_13_camera 目录中。

```
pi@raspberrypi:~ $ cp -r pro_13_camera_source_template pro_13_camera
```

3. 修改 dji_init.c

（1）填写用户账号信息。

```
#define USER_App_NAME              "RaspberryPi"
#define USER_App_ID                "126413"
```

```
#define USER_App_KEY                    "0ea2eadb314d1a6e870975364e2bf4a"
#define USER_App_LICENSE                "HZ/GptycWQ........................"
#define USER_DEVELOPER_ACCOUNT          "abc@outlo**.com"
#define USER_BAUD_RATE                  "460800"
```

（2）开启网络。

```
/*--在通信参数中开启串口和网络功能,定义 CONFIG_HARDWARE_CONNECTION 为
DJI_USE_UART_AND_NETWORK_DEVICE*/
/*--------定义通信参数--------*/
#define DJI_USE_ONLY_UART                       (0)
#define DJI_USE_UART_AND_USB_BULK_DEVICE        (1)
#define DJI_USE_UART_AND_NETWORK_DEVICE         (2)

#define CONFIG_HARDWARE_CONNECTION      DJI_USE_UART_AND_NETWORK_DEVICE
/* -------------------------------------------------------------------*/
```

4. 修改 CMakeLists.txt

在 CMakeLists.txt 中需要设置项目的工程名、添加相机基础功能的目录和文件，并添加树莓派相机需要用到的库文件。按照提示找到相应的位置进行修改即可。

（1）设置本项目的工程名为 pro_13_camera。

```
# -------------------------设置工程名-------------------------
project(pro_13_camera C)
# -----------------------------------------------------------
```

（2）添加目录、文件和库文件。

```
# step 1 添加相机基础功能源文件 camera_emu/*.c 和目录 ./camera_emu ----
file(GLOB_RECURSE MODULE_SAMPLE_SRC camera_emu/*.c)
include_directories(./camera_emu)

# step 2 添加树莓派头文件目录 ./raspi_camera/include --------------
include_directories(./raspi_camera/include)

# step 3 添加连接目录 raspi_camera/lib
link_directories(raspi_camera/lib)

# step 4 添加树莓派连接库
${CMAKE_CURRENT_LIST_DIR}/raspi_camera/lib/libRaspiCamera.a
link_libraries(${CMAKE_CURRENT_LIST_DIR}/raspi_camera/lib/libRaspiCamera.a)
# -----------------------------------------------------------
```

5. 修改 main.c

（1）添加头文件。

```
/* --在此处添加相机基础功能头文件------------------------------------*/
#include "camera_emu/payload_cam_emu_base.h"
/*--------------------------------------------------------------*/
```

（2）添加相机功能和媒体功能服务入口函数。

```
int main(int argc, char **argv)
```

```c
{
    /* ----------此处完成初始化相机功能----------------------------*/

    /* 调用 DjiTest_CameraEmuBaseStartService 函数初始化相机功能---*/
    returnCode = DjiTest_CameraEmuBaseStartService();
    if (returnCode != DJI_ERROR_SYSTEM_MODULE_CODE_SUCCESS) {
        USER_LOG_ERROR("camera emu common init error");
    }
}
```

6. 修改 payload_cam_emu_base.h

相机基础功能头文件 payload_cam_emu_base.h 无须修改。

7. 修改 payload_cam_emu_base.c

根据 payload_cam_emu_base.c 的模板文件,需要完成以下两个函数的功能。

(1) 完成服务入口函数。

在服务入口函数 DjiTest_CameraEmuBaseStartService 中完成以下功能。

```c
/* 在相机基础功能服务入口函数 DjiTest_CameraEmuBaseStartService 中按照提示完成以下功能
--------*/
T_DjiReturnCode DjiTest_CameraEmuBaseStartService(void)
{
    /* step 1 调用函数 DjiPayloadCamera_Init 初始化相机类负载设备 ------*/
    returnCode = DjiPayloadCamera_Init();
    if (returnCode != DJI_ERROR_SYSTEM_MODULE_CODE_SUCCESS) {
        USER_LOG_ERROR("init payload camera error:0x%08llX", returnCode);
        return returnCode;
    }
    /* ---------------------------------------------------------------*/


    /* step 2 在结构体 T_DjiCameraCommonHandler 对象 s_commonHandler 中添加拍照和录像
功能*/
    s_commonHandler.StartShootPhoto = StartShootPhoto;
    s_commonHandler.StopShootPhoto = StopShootPhoto;
    s_commonHandler.StartRecordVideo = StartRecordVideo;
    s_commonHandler.StopRecordVideo = StopRecordVideo;
    /* ---------------------------------------------------------------*/


    /* step 3 调用 DjiPayloadCamera_RegCommonHandler 函数注册相机基础功能-*/
    returnCode = DjiPayloadCamera_RegCommonHandler(&s_commonHandler);
    if (returnCode != DJI_ERROR_SYSTEM_MODULE_CODE_SUCCESS) {
        USER_LOG_ERROR("camera register common handler error:0x%08llX", returnCode);
    }
    /* ---------------------------------------------------------------*/
```

（2）完成拍照函数。

在拍照函数 StartShootPhoto 下完成拍照功能。

```
static T_DjiReturnCode StartShootPhoto(void)
{
    /* ------------------------------添加拍照功能--------------------*/
    if (s_cameraShootPhotoMode == DJI_CAMERA_SHOOT_PHOTO_MODE_SINGLE) {
        s_cameraState.shootingState = DJI_CAMERA_SHOOTING_SINGLE_PHOTO;
        snprintf(cmdStr, sizeof(cmdStr), "sudo raspistill -o a%d.jpg -w 640 -h 480",s_photonum);
        s_photonum++;
        DjiUserUtil_RunSystemCmd(cmdStr);
    /* ----------------------------------------------------------*/
}
```

（3）完成录像函数。

在录像函数 StartRecordVideo 下完成录像功能。

```
static T_DjiReturnCode StartRecordVideo(void)
{
    /* ---添加录像功能-------------------------------------------*/
    snprintf(cmdStr, sizeof(cmdStr), "sudo raspivid -o b%d.h264 -w 1280 -h 720",s_vedionum);
    s_vedionum++;
    DjiUserUtil_RunSystemCmd(cmdStr);
    /*----------------------------------------------------------*/
}
```

8. 编译并执行项目

```
# 在项目 pro_13_camera 目录下，创建 build 目录，并进入 build 目录
pi@raspberrypi:~/pro_13_camera$ mkdir build && cd build

# 使用 cmake 创建项目编译环境
pi@raspberrypi:~/pro_13_camera/build $ cmake ..

# make 编译
pi@raspberrypi:~/pro_13_camera/build $ make

# 执行文件位于 build/bin 目录下
pi@raspberrypi:~/pro_13_camera/build $ cd bin/

# 使用 sudo 执行项目的执行文件
pi@raspberrypi:~/pro_13_camera/build/bin $ sudo ./pro_13_camera
```

任务评价

任务过程评价表

任务实施人姓名_____ 学号_____ 时间_____

评价项目及标准		分值	小组评议	教师评议
技术能力	1. 对基本概念的熟悉程度	10		
	2. 掌握相机基础功能	10		
	3. 掌握拍照功能的实现方法	10		
	4. 掌握照片的存储方法	10		
	5. 掌握录像功能的实现方法	10		
	6. 掌握相机拍照和录像功能的测试方法	10		
执行能力	1. 出勤情况	5		
	2. 遵守纪律情况	5		
	3. 是否主动参与，有无提问记录	5		
	4. 有无职业意识	5		
社会能力	1. 能否有效沟通	5		
	2. 是否使用基本的文明礼貌用语	5		
	3. 能否与组员主动交流、积极合作	5		
	4. 能否自我学习、自我管理	5		
总分		100		
评定等级：				
评价意见		学习意见		

评定等级具体如下。

A：优，得分>90。B：好，90≥得分>80。C：一般，80≥得分>60。D：有待提高，得分≤60。

小结

本项目简要介绍了使用 PSDK 提供的相机功能，通过负载设备（树莓派）上的摄像头实现拍照和录像等相机基础功能的方法，以及云台功能的基本概念。

第 3 篇
PSDK 综合应用

- 项目 14　喊话器
- 项目 15　空气质量检测仪
- 项目 16　AI 目标识别

项目 14 喊话器

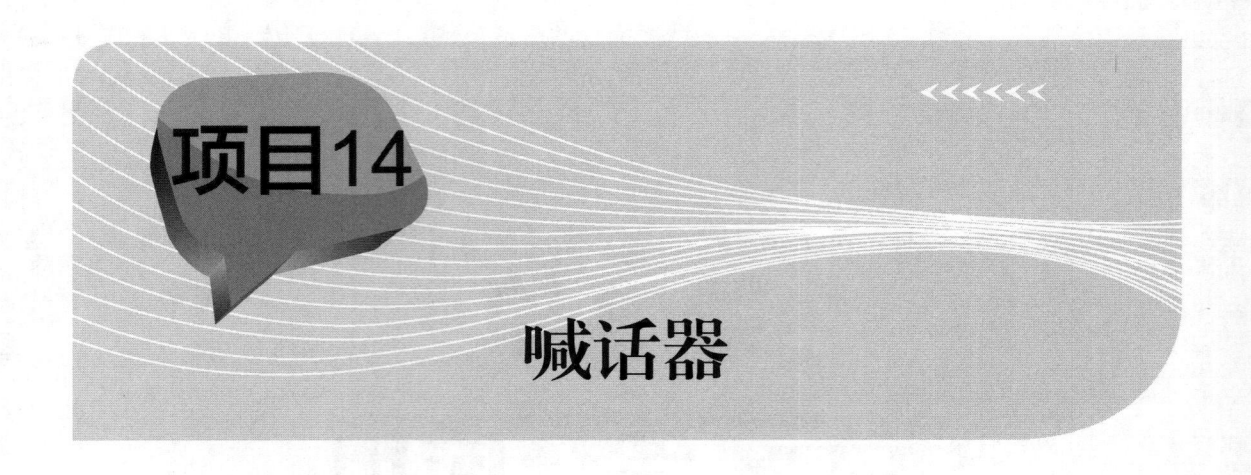

➡ 任务要求

本项目使用 PSDK 提供的喊话器功能,通过负载设备实现实时喊话、文字转语音喊话(TTS)等功能。

➡ 知识导入

14.1 音频编码基础

DJI PSDK 为了开发者能够开发出满足各种搜救、巡逻场景需求的喊话器负载,提供了标准的喊话器解决方案,包括实时喊话、文字转语音喊话(TTS)、音量控制和播放模式控制等功能。开发者无须额外开发 App,就可以直接通过 DJI Pilot 进行喊话器集成开发。

这里只介绍本项目涉及的音频编码基础知识。

1. Opus 音频编解码器

Opus 是一款完全开放、免版税、功能多样的音频编解码器。它适用于互联网上的交互式语音和音乐传输,也适用于存储和流媒体应用。

Opus 的前身是 celt 编码器。在有损音频格式争夺战中,拥有众多不同编码器的 AAC 格式打败了同样颇有潜力的 Musepack、Vorbis 等格式,而在 Opus 格式诞生后,情况似乎发生了变化。通过诸多的对比测试,低码率下的 Opus 完胜曾经优势明显的 HE AAC,其在中码率下就已经可以匹敌码率高出 30%左右的 AAC 格式,而在高码率下更接近原始音频。

Opus 可以处理各种音频应用,包括 IP 语音、视频会议、游戏内聊天,甚至远程现场音乐表演。它可以从低比特率窄带语音扩展到非常高质量的立体声音乐。支持的特性如下。
- 比特率从 6kb/s 到 510 kb/s。
- 采样率从 8kHz（窄带）到 48kHz（全频段）。
- 帧大小从 2.5ms 到 60ms。
- 支持恒定比特率（CBR）和可变比特率（VBR）。
- 从窄带到全频带的音频带宽。
- 支持语音和音乐。
- 支持单声道和立体声。
- 支持多达 255 个通道（多流帧）。
- 动态可调比特率、音频带宽和帧大小。
- 良好的稳健性和隐蔽性。
- 浮点和定点实现。

2. PCM 数据

音频的原始数据是脉冲编码调制（Pulse Code Modulation，PCM）数据，是对连续变化的模拟信号进行抽样、量化和编码而产生的数字信号。描述一段 PCM 数据一般需要 3 个概念：量化格式、采样率（Sample Rate）、声道数（Channel）。

3. 采样率

音频采样率是指录音设备一秒钟内对声音信号的采样次数，采样率越高，还原的声音就越真实、越自然。在当今的主流采集卡中，采样率一般分为 11025Hz、22050Hz、24000Hz、44100Hz、48000Hz 五个等级。

频率对应时间轴线，振幅对应电平轴线。波是无限光滑的，弦线可以看成由无数个点组成，由于存储空间是相对有限的，因此在数字编码过程中，必须对弦线的点进行采样。采样的过程就是抽取某个点的频率值，很显然，一秒钟内抽取的点越多，获取的频率信息就越丰富。为了复原波形，在一次振动中，必须有两个点的采样。人耳能够感觉到的最高频率为 20kHz，因此要想满足人耳的听觉要求，则每秒至少需要进行 40k 次采样，使用 40kHz 表达，这个 40kHz 就是采样率。例如，我们常见的 CD，其采样率为 44.1kHz。

对于挂载在无人机上的喊话器负载，它需要通过地面端的遥控器对音频进行编码，并将编码后的音频上传至无人机端，为了最大化传输效率，需要在音质和采样率之间进行权衡。

4. 声道

声道是指声音在录制或播放时在不同空间位置采集或回放的相互独立的音频信号，所以声道数也就是声音录制时的音源数量或播放时相应的扬声器数量。

对于挂载在无人机上的喊话器负载，受无人机载重和功率的限制，一般选择单扬声器方案，即单通道。

5. TTS

TTS 是 Text To Speech 的缩写，即"从文本到语音"，是人机对话的一部分，也是让机器能够说话的一种文本转语音技术。目前 TTS 技术发展较为成熟，主要包括两类：一类是通过专用的 TTS 硬件芯片进行音频合成，另一类是通过在线或离线的软件库进行 TTS 转换合成，如讯飞离线语音合成 SDK。

14.2 ekho（余音）

ekho（余音）是一个免费、开源的中文语音合成软件，是可以将文字转换成声音的软件。它目前支持粤语、普通话、诏安客语、藏语、雅言（中国古代通用语）和韩语（试验中），英文则通过 Festival 间接实现。更多信息请参阅 ekho 官网。

下面是在树莓派上安装 ekho-6.3.2 的命令，过程如下。

1. 安装依赖库

```
sudo apt-get install libsndfile1-dev libpulse-dev libncurses5-dev libmp3lame-dev
sudo apt-get install libespeak-dev libgtk2.0-dev libvorbis-dev
sudo apt-get install libestools2.1-dev festival-dev
```

2. 解压缩并安装

```
sudo tar -xvf ekho-6.3.2.tar.xz
cd ekho-6.3.2/
sudo ./configure --enable-festival
sudo make CXXFLAGS=-DNO_SSE
sudo make install
```

3. 测试

如果直接输入"ekho '你好'"命令会报错，则建议先将文字转换成语音文件再播放。

```
echo '你好' > 1.txt
ekho -f 1.txt -o 2.wav
ffplay 2.wav

ekho -f 1.txt -o 2.mp3
ffplay 2.mp3
```

14.3 喊话器控件

DJI Pilot 2 支持显示标准喊话器控件，用户可以通过配置自定义控件中的喊话器 json 字段来控制是否进行图标显示。注意：在使用喊话器控件之前，需要先使能自定义控件。

1. 语音喊话

用户可以通过"点击说话"按钮，录制需要喊话的内容。在录制过程中，Pilot 端会有时间计时，当即将达到最长录制时间时，遥控器会震动并进行其他交互提示。当录制完成后，用户点击"结束"按钮，可以选择"本地试听"或"喊话"的方式进行播放。此外，当录制效果不理想时，可以重录，也可以保存当次录制的音频文件，并通过音频列表对其进行管理（支持试听、文件信息显示、删除、重命名、上传播放等）。

2. TTS 喊话

用户可以在控件的文本输入框中直接输入需要喊话的文本，DJI Pilot 2 会将文本内容直接传输到喊话器端，而喊话器端可以结合自身情况自行实现 TTS 功能。

14.4 使用喊话器功能

1. 初始化控件模块

对于喊话器控件的显示，可以通过 speaker 选项中的 json 字段来控制 DJI Pilot 端喊话器 TTS 图标和语音喊话图标的显示。说明：使用喊话器功能前需要使能 Widget 功能。

```
"main_interface": {
  "floating_window": {
    "is_enable": true
  },
  "speaker": {
    "is_enable_tts": true,
    "is_enable_voice": true
  },
```

2. 注册喊话器回调函数

为了实现喊话器功能，需要调用 DjiWidget_RegSpeakerHandler 函数注册喊话器具体功能的实现函数。

```
T_DjiReturnCode DjiTest_WidgetSpeakerStartService(void)
{
    T_DjiReturnCode returnCode;
    T_DjiOsalHandler *osalHandler = DjiPlatform_GetOsalHandler();

    s_speakerHandler.GetSpeakerState = GetSpeakerState;
    s_speakerHandler.SetWorkMode = SetWorkMode;
    s_speakerHandler.StartPlay = StartPlay;
    s_speakerHandler.StopPlay = StopPlay;
    s_speakerHandler.SetPlayMode = SetPlayMode;
    s_speakerHandler.SetVolume = SetVolume;
    s_speakerHandler.ReceiveTtsData = ReceiveTtsData;
```

```
    s_speakerHandler.ReceiveVoiceData = ReceiveAudioData;

    returnCode = DjiWidget_RegSpeakerHandler(&s_speakerHandler);
    if (returnCode != DJI_ERROR_SYSTEM_MODULE_CODE_SUCCESS) {
        USER_LOG_ERROR("Register speaker handler error: 0x%08llX", returnCode);
        return returnCode;
    }
}
```

3. 实现喊话器音频传输

喊话器音频传输需要通过对应的回调函数实现，当喊话器处于 TTS 模式时，在 ReceiveTtsData 回调函数中可以接收到 TTS 文本数据；当喊话器处于实时喊话模式时，在 ReceiveAudioData 回调函数中可以接收到 Opus 音频编码后的数据。在接收数据的过程中，需要经历开始接收、传输数据和完成接收 3 个阶段。

- 开始接收：App 会下发命令到喊话器，喊话器此时准备开始接收音频数据。
- 传输数据：App 会下发音频数据的序号、长度和编码后的数据。
- 完成接收：App 会下发该段音频数据的 MD5 值，并将其提供给喊话器对音频数据进行校验。

```
static T_DjiReturnCode ReceiveTtsData(E_DjiWidgetTransmitDataEvent event,uint32_t offset, uint8_t *buf, uint16_t size)
{
    uint16_t writeLen;
    T_DjiReturnCode returnCode;

    if (event == DJI_WIDGET_TRANSMIT_DATA_EVENT_START) {
        USER_LOG_INFO("Create tts file.");
        s_ttsFile = fopen(WIDGET_SPEAKER_TTS_FILE_NAME, "wb");
        if (s_ttsFile == NULL) {
            USER_LOG_ERROR("Open tts file error.");
        }
        if (s_speakerState.state != DJI_WIDGET_SPEAKER_STATE_PLAYING) {
            SetSpeakerState(DJI_WIDGET_SPEAKER_STATE_TRANSMITTING);
        }
    } else if (event == DJI_WIDGET_TRANSMIT_DATA_EVENT_TRANSMIT) {
        USER_LOG_INFO("Transmit tts file, offset: %d, size: %d", offset, size);
        if (s_ttsFile != NULL) {
            fseek(s_ttsFile, offset, SEEK_SET);
            writeLen = fwrite(buf, 1, size, s_ttsFile);
            if (writeLen != size) {
                USER_LOG_ERROR("Write tts file error %d", writeLen);
            }
        }
        if (s_speakerState.state != DJI_WIDGET_SPEAKER_STATE_PLAYING) {
            SetSpeakerState(DJI_WIDGET_SPEAKER_STATE_TRANSMITTING);
```

```c
        }
    } else if (event == DJI_WIDGET_TRANSMIT_DATA_EVENT_FINISH) {
        USER_LOG_INFO("Close tts file.");
        if (s_ttsFile != NULL) {
            fclose(s_ttsFile);
        }

        returnCode = DjiTest_CheckFileMd5Sum(WIDGET_SPEAKER_TTS_FILE_NAME, buf, size);
        if (returnCode != DJI_ERROR_SYSTEM_MODULE_CODE_SUCCESS) {
            USER_LOG_ERROR("File md5 sum check failed");
        }
        if (s_speakerState.state != DJI_WIDGET_SPEAKER_STATE_PLAYING) {
            SetSpeakerState(DJI_WIDGET_SPEAKER_STATE_IDEL);
        }
    }

    return DJI_ERROR_SYSTEM_MODULE_CODE_SUCCESS;
}
```

4. 实现音频解码

对于编码后的音频数据，需要将其解码成 PCM 格式的 RAW 数据。以下为 Linux 平台的示例代码，需要创建解码器，并按帧读取 Opus 数据，且每帧的大小为 160 字节。将解码后的数据存储为 PCM 格式，以用于播放。

```c
static T_DjiReturnCode DjiTest_DecodeAudioData(void)
{
    //#ifdef OPUS_INSTALLED
    FILE *fin;
    FILE *fout;
    OpusDecoder *decoder;
    opus_int16 out[WIDGET_SPEAKER_AUDIO_OPUS_MAX_FRAME_SIZE *
                   WIDGET_SPEAKER_AUDIO_OPUS_CHANNELS];
    uint8_t cbits[WIDGET_SPEAKER_AUDIO_OPUS_MAX_PACKET_SIZE];
    int32_t nbBytes;
    int32_t err;

    fin = fopen(WIDGET_SPEAKER_AUDIO_OPUS_FILE_NAME, "r");
    if (fin == NULL) {
        fprintf(stderr, "failed to open input file: %s\n", strerror(errno));
        return EXIT_FAILURE;
    }

    /* Create a new decoder state. */
    decoder = opus_decoder_create(WIDGET_SPEAKER_AUDIO_OPUS_SAMPLE_RATE,
```

```c
                                            WIDGET_SPEAKER_AUDIO_OPUS_CHANNELS, &err);
    if (err < 0) {
        fprintf(stderr, "failed to create decoder: %s\n", opus_strerror(err));
        return EXIT_FAILURE;
    }

    fout = fopen(WIDGET_SPEAKER_AUDIO_PCM_FILE_NAME, "w");

    if (fout == NULL) {
        fprintf(stderr, "failed to open output file: %s\n", strerror(errno));

        return EXIT_FAILURE;
    }

    while (1) {
        int i;
        unsigned char pcm_bytes[WIDGET_SPEAKER_AUDIO_OPUS_MAX_FRAME_SIZE
                                * WIDGET_SPEAKER_AUDIO_OPUS_CHANNELS * 2];
        int frame_size;

        /* Read a 16 bits/sample audio frame. */
        nbBytes = fread(cbits, 1, s_decodeBitrate /
                            WIDGET_SPEAKER_AUDIO_OPUS_DECODE_BITRATE_8KBPS *
                        WIDGET_SPEAKER_AUDIO_OPUS_DECODE_FRAME_SIZE_8KBPS, fin);
        if (feof(fin))
            break;
        frame_size = opus_decode(decoder, cbits, nbBytes, out,
                            WIDGET_SPEAKER_AUDIO_OPUS_MAX_FRAME_SIZE, 0);
        if (frame_size < 0) {
            fprintf(stderr, "decoder failed: %s\n", opus_strerror(frame_size));
            return EXIT_FAILURE;
        }

        USER_LOG_DEBUG("decode data to file: %d\r\n", frame_size *
                                    WIDGET_SPEAKER_AUDIO_OPUS_CHANNELS);
        /* Convert to little-endian ordering. */
        for (i = 0; i < WIDGET_SPEAKER_AUDIO_OPUS_CHANNELS * frame_size; i++) {
            pcm_bytes[2 * i] = out[i] & 0xFF;
            pcm_bytes[2 * i + 1] = (out[i] >> 8) & 0xFF;
        }

        /* Write the decoded audio to file. */
        fwrite(pcm_bytes, sizeof(short), frame_size *
                                WIDGET_SPEAKER_AUDIO_OPUS_CHANNELS, fout);
    }
```

```
    /*Destroy the encoder state*/
    opus_decoder_destroy(decoder);
    fclose(fin);
    fclose(fout);

    USER_LOG_INFO("Decode Finished...");
    s_isDecodeFinished = true;

    //#endif
    return EXIT_SUCCESS;
}
```

5. 实现喊话器音频播放

对于喊话器音频的播放，需要结合具体的喊话器负载添加实现方式。Linux 平台系统音频输出默认支持与音频相关的驱动，只需要通过上层软件操作声卡进行音频播放，如 ffplay。

```
static T_DjiReturnCode DjiTest_PlayAudioData(void)
{
    char cmdStr[128];

    memset(cmdStr, 0, sizeof(cmdStr));
    USER_LOG_INFO("Start Playing...");
    snprintf(cmdStr, sizeof(cmdStr), "ffplay -nodisp -autoexit -ar 16000 -ac 1 -f s16le -i %s 2>/dev/null",
             WIDGET_SPEAKER_AUDIO_PCM_FILE_NAME);

    return DjiUserUtil_RunSystemCmd(cmdStr);
}
```

6. 实现 TTS 音频合成

对于喊话器 TTS 音频合成，需要结合具体的喊话器负载添加实现方式。常用的方式有如下两种。

（1）专用 TTS 硬件芯片：该类芯片一般提供 I2C 或 UART 接口的协议，按照协议格式将文本进行转换、发送，即可完成音频的播放。

（2）软件合成库：通过软件库的方式，对输入的文本进行合成，并将其转化成音频数据，该类库对系统资源的占用要求较高，方案较为成熟，如讯飞离线语音合成 SDK。

本项目使用的是开源的 TTS 语言库 ekho。但要想达到更好的 TTS 效果，建议选择其他方式。

```
#if EKHO_INSTALLED
    /*! Attention: you can use other tts opensource function to convert
txt to speech, example used ekho v7.5 */
    snprintf(cmdStr, sizeof(cmdStr), " ekho %s -s 20 -p 20 -a 100 -o %s",
                 data,WIDGET_SPEAKER_TTS_OUTPUT_FILE_NAME);
#else
    USER_LOG_WARN(
```

```
             "Ekho is not installed, please visit https://www.***idedog.net/ekho.php
to install it or use other TTS tools to convert audio");
    #endif
```

14.5 代码模板

在使用喊话器控件之前，需要先使能自定义控件。因此本项目要先实现自定义控件功能，再实现喊话器功能。关于自定义控件功能，参见项目6。

1. 代码结构

下面是实现喊话器功能的代码框架，用户只需要编写用于实现喊话器功能的 Widget_speaker.c 和 Widget_speaker.h 即可。

```
.
├── 3rdparty
│   ├── FindLIBUSB.cmake
│   ├── FindOPUS.cmake              # 这个文件不可少，用于 Opus
│   └── utils
├── CMakeLists.txt
├── dji_init.c
├── dji_init.h
├── hal
├── main.c
├── osal
├── psdk_lib
└── Widget
    ├── file_binary_array_list_en.c
    ├── file_binary_array_list_en.h
    ├── Widget.c                    # 自定义控件源文件
    ├── Widget_file
    ├── Widget_file_c
    ├── Widget.h                    # 自定义控件头文件
    ├── Widget_speaker.c            # 喊话器源代码
    └── Widget_speaker.h            # 喊话器头文件
```

2. 头文件模板

本项目的头文件名为 Widget_speaker.h。该文件定义的内容可满足正常的应用开发需求，如果需要实现特殊的功能，则可以按需要修改。

这里不再展示头文件代码。

3. 源文件模板

本项目的源文件名为 Widget_speaker.c，以下是 Widget_speaker.c 的模板文件，按照提示即可完成代码的编写。

由于源文件的代码行数太多,因此这里仅给出在 Widget_speaker.c 文件中需要完成的功能。

(1) 查找树莓派音频设备名称。

使用 pactl list sinks 命令查找树莓派音频设备名称,如图 14-1 所示。本项目中树莓派使用的音频设备名称为 alsa_output.platform-bcm2835_audio.analog-stereo。需要注意的是,不同音频设备的名称是不一样的。

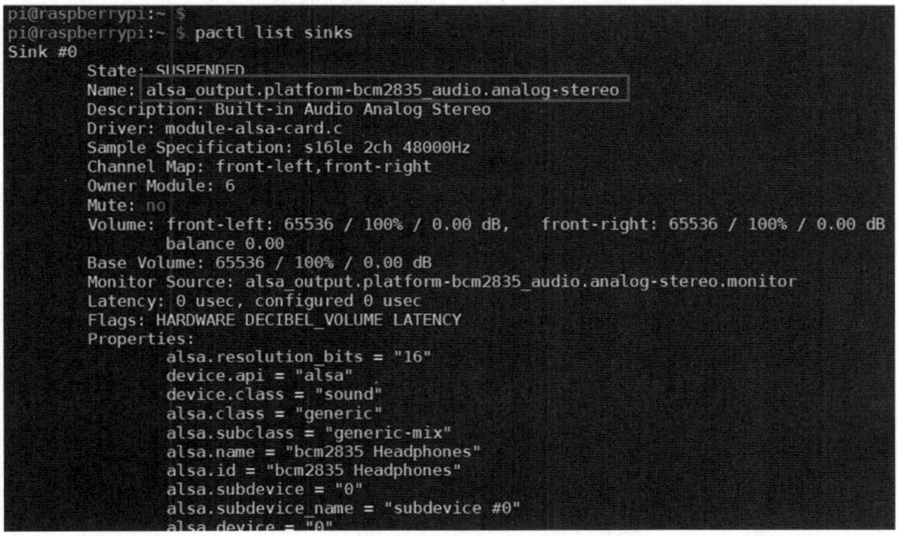

图 14-1 查找树莓派音频设备名称

(2) 在 Widget_speaker.c 中按照提示完成以下功能。

```
/* Includes ------------------------------------------------------------*/
#ifdef SYSTEM_ARCH_LINUX

#include "Widget_speaker.h"
#include "dji_logger.h"
#include <stdlib.h>
#include <errno.h>
#include <string.h>
#include <stdio.h>
#include "utils/util_misc.h"
#include "utils/util_md5.h"
#include <dji_aircraft_info.h>

/* ----------在此处添加OPUSDecoder的头文件 opus.h ------------*/


/* ---------------------------------------------------------------*/
#define WIDGET_SPEAKER_TASK_STACK_SIZE          (2048)
```

```c
/* -------在此处使用 pactl list sinks 命令，添加喊话器使用的音频设备名称----*/
#define WIDGET_SPEAKER_USB_AUDIO_DEVICE_NAME    "换成音频设备名称"

/* ----------------------------------------------------------------*/

/* 在服务入口函数 DjiTest_WidgetSpeakerStartService 中完成以下功能---*/
/* 调用 DjiWidget_RegSpeakerHandler 函数，注册喊话器的实现方法*/

/* ----------------------------------------------------------------*/

T_DjiReturnCode DjiTest_WidgetSpeakerStartService(void)
{
    T_DjiReturnCode returnCode;
    T_DjiOsalHandler *osalHandler = DjiPlatform_GetOsalHandler();

    s_speakerHandler.GetSpeakerState = GetSpeakerState;
    s_speakerHandler.SetWorkMode = SetWorkMode;
    s_speakerHandler.StartPlay = StartPlay;
    s_speakerHandler.StopPlay = StopPlay;
    s_speakerHandler.SetPlayMode = SetPlayMode;
    s_speakerHandler.SetVolume = SetVolume;
    s_speakerHandler.ReceiveTtsData = ReceiveTtsData;
    s_speakerHandler.ReceiveVoiceData = ReceiveAudioData;

    returnCode = osalHandler->MutexCreate(&s_speakerMutex);
    if (returnCode != DJI_ERROR_SYSTEM_MODULE_CODE_SUCCESS) {
        USER_LOG_ERROR("Create speaker mutex error: 0x%08llX", returnCode);
        return returnCode;
    }
    /* step 调用 DjiWidget_RegSpeakerHandler 函数，注册喊话器的实现方法 */

    /* ----------------------------------------------------------------*/
```

任务设计

本项目使用 PSDK 提供的喊话器控件，在负载设备上实现喊话器功能。任务的具体设计要求如下。

（1）从 pro_14_speaker_source_template 项目模板文件中复制一个新项目。文件名为 pro_14_speaker。

（2）本项目将实现喊话器功能，首先需要完成"项目 6 自定义控件"来实现自定义控件，项目目录下 Widget 目录中的 Widget_speaker.c 用于编写代码，以实现喊话器功能。

(3) 在 Widget_speaker.c 中编写代码，实现喊话器功能。
① 安装 ekho 和 Opus 解码器。
② 修改树莓派的音频设备。
③ 注册喊话器功能函数。
(4) 使用 cmake 编译项目。
(5) 执行项目，测试喊话功能。

任务实施

限于篇幅，这里只给出任务实施步骤和关键代码。

1. 模板文件

连接树莓派开发板，默认路径为当前登录用户的家目录。查看当前的家目录，该目录下有一个名为 pro_14_speaker_source_template 的目录，此目录下保存了本项目的所有模板文件。目录结构如下。

```
pi@raspberrypi:~ $ cd pro_14_speaker_source_template/
```

2. 复制模板目录

将 pro_14_speaker_source_template 目录下的所有文件和目录复制到 pro_14_speaker 目录中。

```
pi@raspberrypi:~ $ cp -r pro_14_speaker_source_template pro_14_speaker
```

3. 修改 dji_init.c

填写用户账号信息。

```
#define USER_App_NAME               "RaspberryPi"
#define USER_App_ID                 "126413"
#define USER_App_KEY                "0ea2eadb314d1a6e870975364e2bf4a"
#define USER_App_LICENSE            "HZ/GptycWQ........................"
#define USER_DEVELOPER_ACCOUNT      "abc@outlo**.com"
#define USER_BAUD_RATE              "460800"
```

4. 修改 CMakeLists.txt

在 CMakeLists.txt 中需要设置项目的工程名、添加喊话器功能的目录和文件，并添加 Opus 解码器的库文件。按照提示找到相应的位置进行修改即可。

（1）设置本项目的工程名为 pro_14_speaker。

```
# -------------------------设置工程名-------------------------
project(pro_14_speaker C)
# -----------------------------------------------------------
```

（2）添加目录、文件和库文件。

```
# ------在此处添加 Opus 解码器的 include 目录和 libopus.so 库文件----------

# 添加 Opus 解码器的 include 目录
```

```
include_directories(/usr/include/opus/)

# 添加Opus解码器的libopus.so库文件
link_libraries(${CMAKE_CURRENT_LIST_DIR}/psdk_lib/lib/${TOOLCHAIN_NAME}/libopus.so)

# ------在此处添加Opus解码器的编译信息----------------------------
find_package(OPUS REQUIRED)
if (OPUS_FOUND)
    message(STATUS "Found OPUS installed in the system")
    message(STATUS " - Includes: ${OPUS_INCLUDE_DIR}")
    message(STATUS " - Libraries: ${OPUS_LIBRARY}")
    add_definitions(-DOPUS_INSTALLED)
    target_link_libraries(${PROJECT_NAME}/usr/lib/arm-linux-gnueabihf/libopus.so)
else ()
    message(STATUS "Cannot Find OPUS")
endif (OPUS_FOUND)

# -------------------------------------------------------------------
```

5. 修改 main.c

（1）添加头文件。

```
/* -----在此处添加自定义控件和喊话器头文件----------------------*/
#include "Widget/Widget.h"
#include "Widget/Widget_speaker.h"
/* ---------------------------------------------------------*/
```

（2）添加自定义控件和喊话器功能的服务入口函数。

```
/* --------在main.c中初始化自定义控件和喊话器功能----------------------*/

int main(int argc, char **argv)
{
    /* step 1 调用DjiTest_WidgetStartService函数初始化自定义控件功能----*/
    returnCode = DjiTest_WidgetStartService();
    if (returnCode != DJI_ERROR_SYSTEM_MODULE_CODE_SUCCESS) {
        USER_LOG_ERROR("Widget sample init error");
    }
    /* ---------------------------------------------------------------*/

    /* step 2 调用DjiTest_WidgetSpeakerStartService函数初始化喊话器功能*/
    returnCode = DjiTest_WidgetSpeakerStartService();
    if (returnCode != DJI_ERROR_SYSTEM_MODULE_CODE_SUCCESS) {
        USER_LOG_ERROR("Widget speaker test init error");
    }
    /* ---------------------------------------------------------------*/
}
```

6. 修改 Widget_speaker.h

喊话器头文件 Widget_speaker.h 无须修改。

7. 修改 Widget_speaker.c

根据 Widget_speaker.c 的模板文件，按照代码中的提示完成以下功能。

```c
/* ----------在此处添加OPUSDecoder的头文件opus.h --------------------*/
#include "opus.h"
/* ----------------------------------------------------------------*/


/* -------在此处使用pactl list sinks命令，添加喊话器使用的音频设备名-------*/
#define WIDGET_SPEAKER_USB_AUDIO_DEVICE_NAME     "alsa_output.platform-bcm2835_audio.analog-stereo"

/* -------------------------------------------------------------- */

/* 在服务入口函数 DjiTest_WidgetSpeakerStartService 中完成以下功能----*/

T_DjiReturnCode DjiTest_WidgetSpeakerStartService(void)
{
    /* 调用 DjiWidget_RegSpeakerHandler 函数，注册喊话器的实现方法 */
    returnCode = DjiWidget_RegSpeakerHandler(&s_speakerHandler);
    if (returnCode != DJI_ERROR_SYSTEM_MODULE_CODE_SUCCESS) {
        USER_LOG_ERROR("Register speaker handler error: 0x%08llX", returnCode);
        return returnCode;
    }
    /* --------------------------------------------------------------*/
```

8. 编译并执行项目

```
# 在项目pro_14_speaker目录下，创建build目录，并进入build目录
pi@raspberrypi:~/pro_14_speaker$ mkdir build && cd build

# 使用cmake创建项目编译环境
pi@raspberrypi:~/pro_14_speaker/build $ cmake ..

# make 编译
pi@raspberrypi:~/pro_14_speaker/build $ make

# 执行文件位于build/bin目录下
pi@raspberrypi:~/pro_14_speaker/build $ cd bin/

# 使用sudo执行项目的执行文件
pi@raspberrypi:~/pro_14_speaker/build/bin $ sudo ./pro_14_speaker
```

任务评价

任务过程评价表

任务实施人姓名_____ 学号_____ 时间_____

评价项目及标准		分值	小组评议	教师评议
技术能力	1. 对基本概念的熟悉程度	10		
	2. 掌握 ekho 的安装及使用方法	10		
	3. 掌握 Opus 编码器的用途	10		
	4. 掌握 ekho TTS 的使用方法	10		
	5. 掌握查看音频设备的方法	10		
	6. 掌握喊话器功能的测试方法	10		
执行能力	1. 出勤情况	5		
	2. 遵守纪律情况	5		
	3. 是否主动参与，有无提问记录	5		
	4. 有无职业意识	5		
社会能力	1. 能否有效沟通	5		
	2. 是否使用基本的文明礼貌用语	5		
	3. 能否与组员主动交流、积极合作	5		
	4. 能否自我学习、自我管理	5		
总分		100		
评定等级：				
评价意见		学习意见		

评定等级具体如下。

A：优，得分>90。B：好，90≥得分>80。C：一般，80≥得分>60。D：有待提高，得分≤60。

小结

本项目首先简要介绍了喊话器功能的基本概念和使用方法；其次介绍了 ekho 和 Opus 的基本概念与基本功能；最后介绍了实现喊话器功能的代码框架。

项目15 空气质量检测仪

任务要求

本项目通过自定义控件的方式实现简单的空气质量检测仪。

知识导入

本项目将基于自定义控件来实现,关于自定义控件的相关知识,参见项目 6。

15.1 硬件设计

硬件控制部分主要由 Raspberry Pi 4B、MQ-135 空气质量传感器、MCP3008 A/D 信号转换器、面包板等组件组成。

1. MQ-135 空气质量传感器

MQ-135 空气质量传感器(也称气体传感器)所使用的气敏材料是在清洁空气中电导率较低的二氧化锡(SnO_2)。当传感器所处环境中存在污染气体时,传感器的电导率会随空气中污染气体浓度的增加而增大。MQ-135 空气质量传感器对氨气、硫化物、苯系蒸气的灵敏度高,对烟雾和其他有害气体的监测效果也很理想。这种传感器可检测多种有害气体,是一款适合多种应用场景的低成本传感器。

MQ-135 空气质量传感器主要用于家庭和环境中的有害气体检测,如图 15-1 所示。

图 15-1 MQ-135 空气质量传感器

2. MCP3008 A/D 信号转换器

由于树莓派开发板上没有模拟信号的输入引脚，因此树莓派不能直接处理模拟输入信号。为了解决这个问题，本项目通过使用 MCP3008 A/D 信号转换器来解决树莓派模拟信号输入的问题。

MCP3008 是一个 8 通道的 10 位 ADC 芯片，具有 10 位分辨率来测量 8 个不同的模拟电压，输出最大值为 1023，所以输出的范围是 0~1023。MCP3008 通过 SPI 通信将该值发送到微控制器或微处理器中。MCP3008 芯片可以在 3.3V 和 5V 的电压下工作，还可以与 5V 的微控制器及 3.3V 的系统（如 Raspberry Pi）一起使用。

MCP3008 芯片本身在一端标有一个缺口，代表引脚 1 和 16。MCP3008 芯片引脚如图 15-2 所示。

MCP3008 芯片共有 16 个引脚，其中，左侧 8 个引脚用于接收模拟信号输入，分别是 CH0~CH7（引脚序号：1~8）。D.Ground 是芯片的数字地，CS'/SHDN 是芯片选择引脚，D$_{IN}$ 和 D$_{OUT}$ 分别是数据输入和输出脚，Clock 是时钟，A.Ground 是模拟地，V$_{REF}$ 是模拟基准电压脚，Vdd/Vcc 是供电引脚。

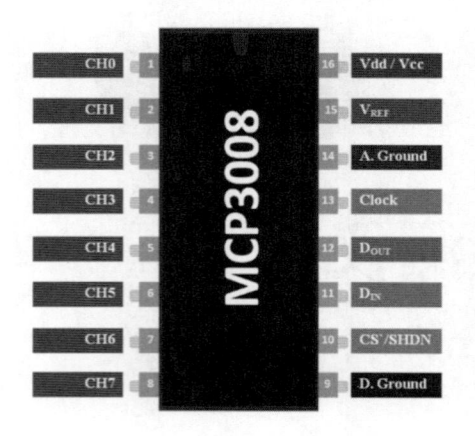

图 15-2　MCP3008 芯片引脚

3. 使能树莓派的 SPI 接口

打开树莓派的命令终端并输入 sudo raspi-config 命令，弹出树莓派配置界面。选择"3 Interface Options　Configure connections to peripherals"选项，如图 15-3 所示。

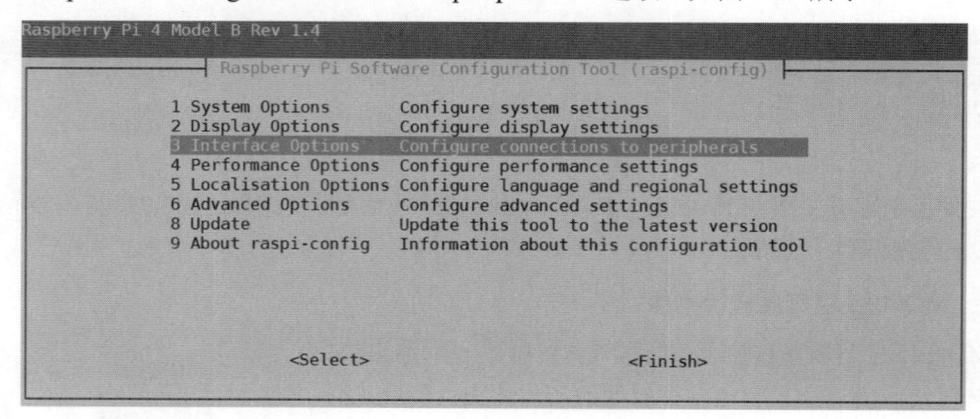

图 15-3　选择接口选项

在跳转的界面中选择"I4 SPI　Enable/disable automatic loading of SPI kernel module"选项。使能 SPI，如图 15-4 所示。

4. 面包板

采用面包板连接线路，面包板如图 15-5 所示。

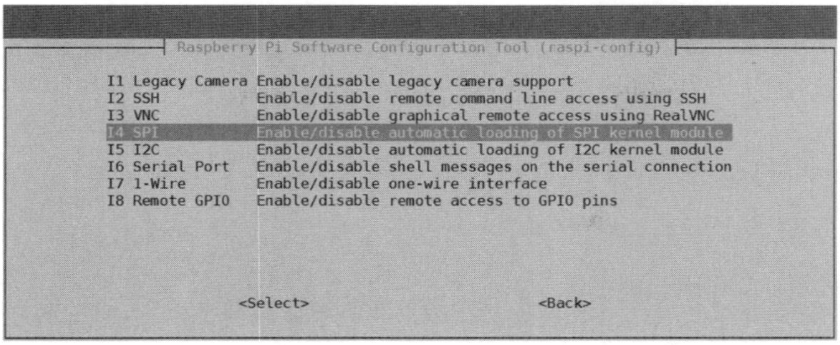

图 15-4 使能 SPI

5. 连接方式

整体连接方式如图 15-6 所示。

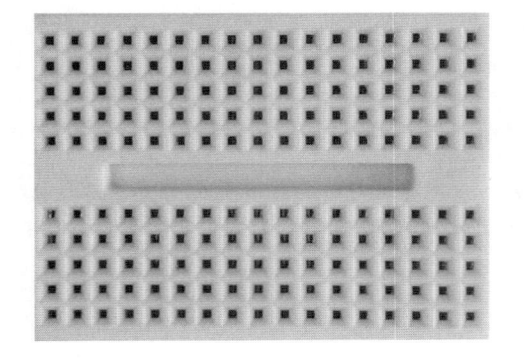

图 15-5 面包板 　　　　　　　　　　　图 15-6 整体连接方式

15.2 模数转换

下面编写了一个使用 C++实现 MCP3008 模数转换的代码，其中包括两个文件，一个头文件 MCP3008.h 和一个源代码文件 MCP3008.cpp。详细过程描述如下。

（1）MCP3008 读取 MQ-135 空气质量传感器的模拟信号。
（2）MCP3008 将模拟信号转换为数字信号，并将其保存在缓冲区中。
（3）树莓派通过 SPI 读取 MCP3008 保存在缓冲区中的数字信号，并进行下一步处理。

1. 头文件 MCP3008.h

```
#ifndef MCP3008_H_AA185758_F169_4B8A_8158_6E4588F5B55F
#define MCP3008_H_AA185758_F169_4B8A_8158_6E4588F5B55F

#include <stdint.h>
#include <linux/spi/spidev.h>
```

```cpp
namespace MCP3008Lib {

enum class Mode{
    SINGLE = 1,
    DIFFERENTIAL = 0
};

class MCP3008 {
public:
    static const int DEFAULT_SPI_DEV = 0;
    static const int DEFAULT_SPI_CHANNEL = 0;
    static const int SPI_5V_BAUD = 3600000;
    static const int SPI_2_7V_BAUD = 1350000;
    static const int DEFAULT_SPI_BAUD = SPI_2_7V_BAUD;
    static const int DEFAULT_SPI_FLAGS = SPI_MODE_0;

    MCP3008(
        const int dev = DEFAULT_SPI_DEV,
        const int channel = DEFAULT_SPI_CHANNEL,
        const int baud = DEFAULT_SPI_BAUD,
        const int flags = DEFAULT_SPI_FLAGS) noexcept;

    virtual ~MCP3008();

    void connect();
    void disconnect();
    unsigned short read(const uint8_t channel, const Mode m = Mode::SINGLE) const;

protected:
    int _handle;
    int _dev;
    int _channel;
    int _baud;
    int _flags;

};

};
#endif
```

2. 源代码文件 MCP3008.cpp

```cpp
#include "../include/MCP3008.h"
#include <cstdint>
#include <lgpio.h>
#include <stdexcept>
```

```cpp
namespace MCP3008Lib {

MCP3008::MCP3008(
    const int dev,
    const int channel,
    const int baud,
    const int flags) noexcept :
        _handle(-1),
        _dev(dev),
        _channel(channel),
        _baud(baud),
        _flags(flags) {
}

MCP3008::~MCP3008() {

    try {
        this->disconnect();
    }
    catch(...) {
        //prevent propagation
    }

}

void MCP3008::connect() {

    if(this->_handle >= 0) {
        return;
    }

    const auto handle = ::lgSpiOpen(
        this->_dev,
        this->_channel,
        this->_baud,
        this->_flags);

    if(handle < 0) {
        throw std::runtime_error("failed to connect spi device");
    }

    this->_handle = handle;

}

void MCP3008::disconnect() {
```

```cpp
        if(this->_handle < 0) {
            return;
        }

        if(::lgSpiClose(this->_handle) != 0) {
            throw std::runtime_error("failed to disconnect spi device");
        }

        this->_handle = -1;

}

unsigned short MCP3008::read(const std::uint8_t channel, const Mode m) const {

    //control bits
    //first bit is single or differential mode
    //next three bits are channel selection
    //last four bits are ignored
    const std::uint8_t ctrl =
        (static_cast<std::uint8_t>(m) << 7) |
         static_cast<std::uint8_t>((channel & 0b00000111) << 4)
        ;

    const std::uint8_t byteCount = 3;

    const std::uint8_t txData[byteCount] = {
        0b00000001, //seven leading zeros and start bit
        ctrl,       //sgl/diff (mode), d2, d1, d0, 4x "don't care" bits
        0b00000000  //8x "don't care" bits
        };

    std::uint8_t rxData[byteCount]{0};

    const auto bytesTransferred = ::lgSpiXfer(
        this->_handle,
        reinterpret_cast<const char*>(txData),
        reinterpret_cast<char*>(rxData),
        byteCount);

    if(bytesTransferred != byteCount) {
        throw std::runtime_error("spi transfer failed");
    }

    //first 14 bits are ignored
    //no need to AND with 0x3ff this way
```

```
    return
        ((static_cast<unsigned short>(rxData[1]) & 0b00000011) << 8) |
         (static_cast<unsigned short>(rxData[2]) & 0b11111111);
}

};
```

任务设计

本项目通过自定义控件的方式实现简单的空气质量检测仪,具体设计要求如下。
(1)硬件连接。
① 在面包板上安装 MQ-135 空气质量传感器。
② 在面包板上安装 MCP3008 A/D 信号转换器。
③ 连接树莓派 SPI 到面包板。
(2)使能树莓派 SPI 串口。
(3)完成项目代码的编写。
(4)执行项目。

任务实施

限于篇幅,这里只给出任务实施步骤和关键代码。

1. 构建项目

连接树莓派开发板,默认路径为当前登录用户的家目录。本项目将基于项目 6 实现,将已经完成的项目 6 的项目文件复制到本项目中,项目名为 pro_15_aircheck。

```
pi@raspberrypi:~ $ cp -r pro_6_Widget pro_15_aircheck
```

2. 在本项目的目录下创建一个目录

(1)添加 aircheck 目录。

在 pro_15_aircheck 目录下按照下面的目录结构创建 aircheck 目录。其中,MCP3008.h 和 MCP3008.cpp 参见 15.2 节。

```
├── aircheck
│   ├── include
│   │   └── MCP3008.h
│   └── src
│       └── MCP3008.cpp
```

(2)将 Widget/Widget.c 修改成 Widget/Widget.cpp。

3. 修改 dji_init.c

填写用户账号信息。

```
#define USER_App_NAME            "RaspberryPi"
#define USER_App_ID              "126413"
```

```
#define USER_App_KEY                "0ea2eadb314d1a6e870975364e2bf4a"
#define USER_App_LICENSE            "HZ/GptycWQ........................"
#define USER_DEVELOPER_ACCOUNT      "abc@outlo**.com"
#define USER_BAUD_RATE              "460800"
```

4. 修改 CMakeLists.txt

（1）设置本项目的工程名为 pro_15_aircheck，并添加 CXX 选项。

```
# -------------------------设置工程名-------------------------
project(pro_15_aircheck CXX C)
# ----------------------------------------------------------
```

（2）在 CMAKE_CXX_FLAGS 中添加 -Wall -Wextra 选项。

```
# -----在 CMAKE_CXX_FLAGS 中添加 -Wall -Wextra 选项--------------------
set(CMAKE_CXX_FLAGS "${CMAKE_CXX_FLAGS} -fprofile-arcs -ftest-coverage -Wall -Wextra")
# ----------------------------------------------------------
```

（3）添加目录和文件。

```
# ------在此处添加 Widget 和 aircheck 的 /*.cpp 和目录------------------
file(GLOB_RECURSE MODULE_AIRCHECK_SRC aircheck/src/*.cpp)
file(GLOB_RECURSE MODULE_SAMPLE_SRC Widget/*.cpp)
include_directories(./Widget)
include_directories(./aircheck/include)
# ----------------------------------------------------------
```

（4）添加 GPIO 库文件。

```
link_libraries(/usr/local/lib/liblgpio.so)
```

（5）添加 MODULE_AIRCHECK_SRC 到 add_executable 中。

```
add_executable(${PROJECT_NAME}
        main.c
        dji_init.c
        ${MODULE_OSAL_SRC}
        ${MODULE_HAL_SRC}
        ${MODULE_3RDPART_SRC}
        ${MODULE_SAMPLE_SRC}
        ${MODULE_AIRCHECK_SRC})      # 添加 MODULE_AIRCHECK_SRC
```

5. 修改 main.c

main.c 不需要修改，直接使用项目 6 的 main.c 即可。

6. 修改 MCP3008.h

MCP3008.h 无须修改。

7. 修改 MCP3008.cpp

注意：这里文件后缀名为 .cpp。

（1）添加头文件。
```
/*----在此处添加 MCP3008.h----------------*/
#include "MCP3008.h"
/* ---------------------------------------*/
```
（2）在工作线程（任务）DjiTest_WidgetTask 中添加以下代码。
```
static void *DjiTest_WidgetTask(void *arg)
{

USER_UTIL_UNUSED(arg);

/* --调用 MCP3008 的函数，读取数字信号----------*/
    MCP3008Lib::MCP3008 adc;
adc.connect();
/* ---------------------------------------/*

    while (1) {

        /* 保留源代码：调用 GetTimeMs 函数获取当前系统时间--------*/
        djiStat = osalHandler->GetTimeMs(&sysTimeMs);
        if (djiStat != DJI_ERROR_SYSTEM_MODULE_CODE_SUCCESS) {
            USER_LOG_ERROR("Get system time ms error, stat = 0x%08llX", djiStat);
        }
        /* ---------------------------------------------/*

        /* --调用 snprintf 函数拼接输出消息，读取 MCP3008 的 CH0 的数据----*/
        snprintf(message, DJI_WIDGET_FLOATING_WINDOW_MSG_MAX_LEN, "System time : %u ms\nair check : %d", sysTimeMs, adc.read(0));
        /* ---------------------------------------------------*/

        /*---向 DJI Pilot 的浮窗发送消息-------------*/
        djiStat = DjiWidgetFloatingWindow_ShowMessage(message);
        if (djiStat != DJI_ERROR_SYSTEM_MODULE_CODE_SUCCESS) {
            USER_LOG_ERROR("Floating window show message error, stat = 0x%08llX", djiStat);
        }

        osalHandler->TaskSleepMs(1000);
        /* ------------------------------------------*/

    }
}
```

8. 编译并执行项目

```
# 在项目 pro_15_aircheck 目录下，创建 build 目录，并进入 build 目录
pi@raspberrypi:~/pro_15_aircheck$ mkdir build && cd build
```

```
# 使用cmake创建项目编译环境
pi@raspberrypi:~/pro_15_aircheck/build $ cmake ..

# make 编译
pi@raspberrypi:~/pro_15_aircheck/build $ make

# 执行文件位于build/bin目录下
pi@raspberrypi:~/pro_15_aircheck/build $ cd bin/

# 使用sudo执行项目的执行文件
pi@raspberrypi:~/pro_15_aircheck/build/bin $ sudo ./pro_15_aircheck
```

任务评价

任务过程评价表

任务实施人姓名_____ 学号_____ 时间_____

评价项目及标准		分值	小组评议	教师评议
技术能力	1. 对基本概念的熟悉程度	10		
	2. 了解传感器的简单知识	10		
	3. 了解模数转换的简单知识	10		
	4. 了解树莓派SPI串口	10		
	5. 掌握硬件连接的方法	10		
	6. 掌握测试方法	10		
执行能力	1. 出勤情况	5		
	2. 遵守纪律情况	5		
	3. 是否主动参与，有无提问记录	5		
	4. 有无职业意识	5		
社会能力	1. 能否有效沟通	5		
	2. 是否使用基本的文明礼貌用语	5		
	3. 能否与组员主动交流、积极合作	5		
	4. 能否自我学习、自我管理	5		
总分		100		
评定等级：				
评价意见			学习意见	

评定等级具体如下。

A：优，得分>90。B：好，90≥得分>80。C：一般，80≥得分>60。D：有待提高，得分≤60。

小结

本项目首先简要介绍了空气质量检测仪的硬件设计；其次介绍了模数转换的相关知识；最后介绍了实现简单的空气质量检测仪的方法。

项目16 AI 目标识别

任务要求

本项目基于 M300 上 OSDK 的实时视频流功能来读取相机的实时视频流，并通过 OpenCV 实现人脸检测。

知识导入

16.1 实时视频流功能

本项目基于实时视频流功能实现，关于实时视频流功能的相关知识，请参阅"项目 11 获取相机码流（liveview）"。

项目 11 中使用 C 语言只实现了如何从无人机中读取实时流；而本项目使用 C++实现了基于实时流的解码及图像处理等功能。

16.2 人脸检测

OpenCV 是一种由 C++编写，基于 BSD 许可发行的跨平台计算机视觉与机器学习软件库，可以在 Linux、Windows、Android、macOS 等系统中运行。同时，OpenCV 提供了 C++、C、Python、Java 和 MATLAB 等语言的接口。

目前，OpenCV 已经被广泛应用在多个领域中，人机互动、人脸识别、机器视觉、运动追踪和分析、动作识别、图像分割等领域都可见其身影。它具有强大的跨平台性，属于开源的开发工具，可以免费应用在各领域中。

在实际应用环节中，OpenCV 有着运行速度快、开发目的明确、运行独立性高、图像视频输入/输出效率高且程序底层与高层开发包完善的特点。同时，应用 OpenCV 能够为深度开发计算机视觉市场提供巨大帮助。

应用 OpenCV 可以快速完成图像数据的分配、释放、复制和转换；还可以快速获取文件与摄像头中的图像或视频并完成二者的输出；能够基于奇异值算法、解方程算法或矩阵积算法处理矩阵和向量；还能够进行多元结构分析和数字图像的基本处理，有效开展动物体图像的追踪分析与运动分割。

人脸检测是指在输入图像中确定人脸位置、大小的过程。人脸检测作为人脸信息处理的一项关键技术，近年来成为模式识别与视觉领域中一项受到普遍重视、研究十分活跃的课题。

人脸检测问题来源于人脸识别，是人脸识别系统中的一个必要环节。如果没能检测出人脸，那么人脸识别系统只能是巧妇难为无米之炊。在早期的人脸识别研究中，针对的对象往往是具有较强约束条件的人脸图像，如无背景、固定大小、固定位置的图像，人脸位置已知或很容易获得，因此人脸检测问题并未受到重视。

近年来，随着视觉领域的发展，人脸识别成为具有极大潜力的生物身份验证手段，这要求自动人脸识别系统能够对不同环境下的图像具有稳定的识别能力。因此人脸检测开始作为一个独立的课题受到研究者的重视。在基于内容的检索、数字视频处理、视觉监测等领域中，人脸检测有着重要的应用价值。早期的人脸检测方法大体上可以分为 4 类：基于知识的方法、基于不变性特征的方法、基于模板匹配的方法、基于外观形状的方法。其中，在基于模板匹配的方法中，较为常用的是基于 Haar 特征的人脸检测。

OpenCV 中提供了已经训练完成的可用于人脸检测的模型，安装 OpenCV 后，这些模型默认保存在/usr/local/share/opencv4/haarcascades/目录下，部分模型名称及其检测部位如表 16-1 所示。

表 16-1 部分模型名称及其检测部位

模型名称	检测部位
haarcascade_frontalface_default.xml	默认 Haar 人脸检测
haarcascade_frontalface_alt2.xml	快速 Haar 人脸检测
haarcascade_profileface.xml	侧脸检测
haarcascade_lefteye_2splits.xml	左眼检测
haarcascade_righteye_2splits.xml	右眼检测
haarcascade_fullbody.xml	身体检测
haarcascade_smile.xml	笑脸检测

在 OpenCV 中，通过 CascadeClassifier 函数读取级联分类器，其基本使用格式如下。

```
cv::CascadeClassifier( filename )
```

filename 参数用于接收字符串，即分类器的路径。运行 CascadeClassifier 函数后会返回一个分类器对象，并通过 detectMultiScale 函数实现多个尺度空间上的人脸检测，其基本使用格式如下。

```
cv::CascadeClassifier.detectMultiScale( image[, scaleFactor[, minNeighbors [, minSize[, maxSize]]]]
```

detectMultiScale 函数的参数说明如表 16-2 所示。

表 16-2 detectMultiScale 函数的参数说明

参数名称	说明
image	表示输入的图像，为灰度图。无默认值
scaleFactor	表示尺度变换的比例。默认为 1.1
minNeighbors	表示候选矩形保留相邻矩形的数量。默认为 3
minSize	表示最小的矩形框的大小。无默认值
maxSize	表示最大的矩形框的大小。无默认值

16.3 代码模板

1. 代码结构

下面是实现本项目的代码框架，用户只需要编写 liveview 目录下的 liveview_entry.cpp 和 liveview.hpp 即可。

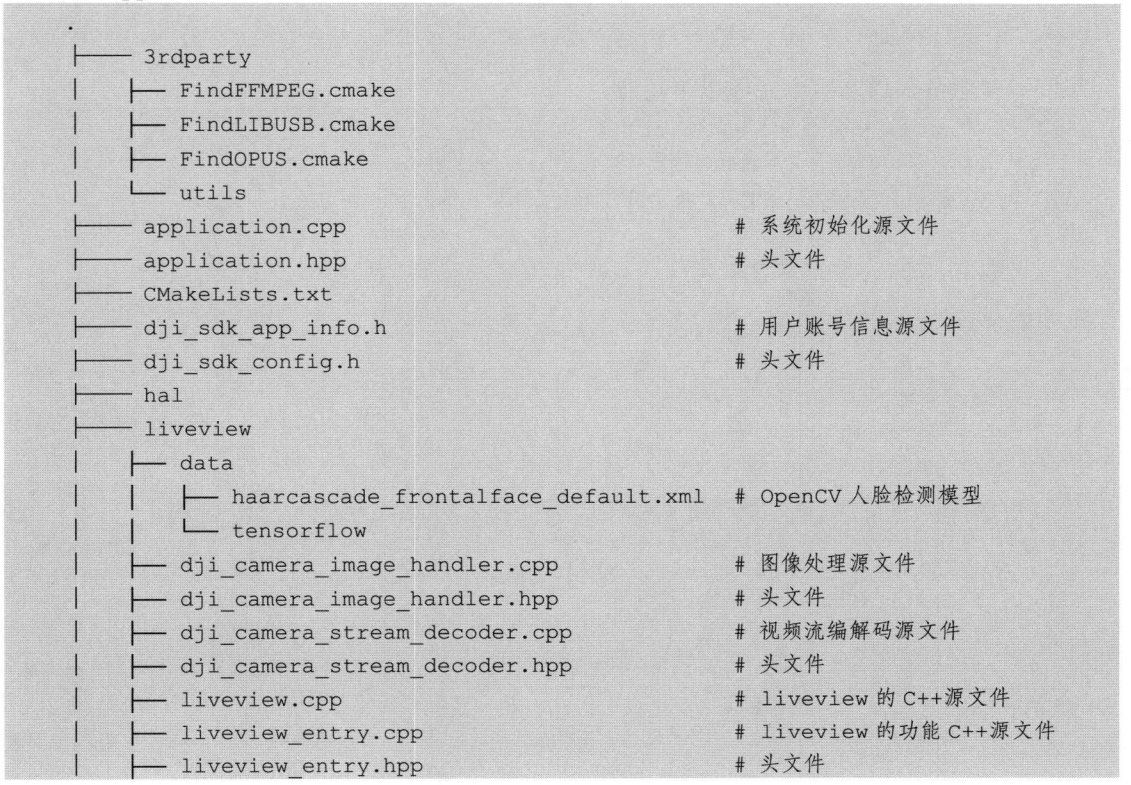

```
.
├── 3rdparty
│   ├── FindFFMPEG.cmake
│   ├── FindLIBUSB.cmake
│   ├── FindOPUS.cmake
│   └── utils
├── application.cpp                           # 系统初始化源文件
├── application.hpp                           # 头文件
├── CMakeLists.txt
├── dji_sdk_app_info.h                        # 用户账号信息源文件
├── dji_sdk_config.h                          # 头文件
├── hal
├── liveview
│   ├── data
│   │   ├── haarcascade_frontalface_default.xml   # OpenCV 人脸检测模型
│   │   └── tensorflow
│   ├── dji_camera_image_handler.cpp          # 图像处理源文件
│   ├── dji_camera_image_handler.hpp          # 头文件
│   ├── dji_camera_stream_decoder.cpp         # 视频流编解码源文件
│   ├── dji_camera_stream_decoder.hpp         # 头文件
│   ├── liveview.cpp                          # liveview 的 C++源文件
│   ├── liveview_entry.cpp                    # liveview 的功能 C++源文件
│   ├── liveview_entry.hpp                    # 头文件
```

```
        │       └── liveview.hpp                    # 头文件
        ├── main.cpp                                # 主函数 C++ 源文件
        ├── osal
        └── psdk_lib
```

2. 头文件模板

头文件不需要修改。

3. 源文件模板

本项目的源文件名为 liveview_entry.cpp，以下是 liveview_entry.cpp 的模板文件，按照提示即可完成代码的编写。

由于源文件的代码行数太多，这里仅给出在 liveview_entry.cpp 中需要完成的功能。

（1）添加 OpenCV 头文件。

```cpp
#include <iostream>
#include <dji_logger.h>
#include "liveview_entry.hpp"
#include "liveview.hpp"
#include "../3rdparty/utils/util_misc.h"

#ifdef OPEN_CV_INSTALLED
/* ----------------添加 OpenCV 头文件------------------ */



/* ---------------------------------------------------- */
```

（2）在服务入口函数中完成以下功能。

```cpp
/* -------------在服务入口函数中完成以下功能---------------*/
/* step 1 启动 FPV 摄像头获取实时视频流---------------------*/
/* step 2 停止 FPV 摄像头获取实时视频流---------------------*/

/* --------------------------------------------------------*/

void DjiUser_RunCameraStreamViewSample()
{
    char cameraIndexChar = 0;
    char demoIndexChar = 0;
    char isQuit = 0;
    CameraRGBImage camImg;
    char fpvName[] = "FPV_CAM";
    auto *liveviewSample = new LiveviewSample();
    T_DjiReturnCode returnCode;

    returnCode = DjiUser_GetCurrentFileDirPath(__FILE__, DJI_FILE_PATH_SIZE_MAX, curFileDirPath);
    if (returnCode != DJI_ERROR_SYSTEM_MODULE_CODE_SUCCESS) {
        USER_LOG_ERROR("Get file current path error, stat = 0x%08llX", returnCode);
```

```
    }
    /* step 1 启动FPV摄像头获取实时视频流---------------------*/


    /* -----------------------------------------------------*/


    cout << "Please enter the 'q' or 'Q' to quit camera stream view\n"
         << endl;

    while (true) {
        cin >> isQuit;
        if (isQuit == 'q' || isQuit == 'Q') {
            break;
        }
    }
    /* step 2 停止FPV摄像头获取实时视频流---------------------*/


    /* -----------------------------------------------------*/


    delete liveviewSample;
}
```

（3）在回调函数中完成以下功能。

```
static void DjiUser_ShowRgbImageCallback(CameraRGBImage img, void *userData)
{
    string name = string(reinterpret_cast<char *>(userData));

#ifdef OPEN_CV_INSTALLED
    printf("-------OPEN_CV_INSTALLED--------\n");
    Mat mat(img.height, img.width, CV_8UC3, img.rawData.data(), img.width * 3);

    cvtColor(mat, mat, COLOR_RGB2BGR);
    snprintf(tempFileDirPath, DJI_FILE_PATH_SIZE_MAX,
"%s/data/haarcascade_frontalface_default.xml", curFileDirPath);

    /* ------在此处完成人脸检测 ---------------------------------*/
    /* step 1 创建级联分类器-----------------------------------*/
    /* step 2 实现人脸检测------------------------------------*/
    /* step 3 将检测的人脸用方框框起来---------------------------*/
    /* step 4 显示检测的人脸----------------------------------*/

    /* step 1 创建级联分类器-----------------------------------*/


    /* -----------------------------------------------------*/


    /* step 2 实现人脸检测------------------------------------*/


    /* -----------------------------------------------------*/
```

```
        /* step 3 将检测的人脸用方框框起来--------------------------*/


        /* --------------------------------------------------------*/

        /* step 4 显示检测的人脸-----------------------------------*/


        /* --------------------------------------------------------*/
        cv::waitKey(1);
#endif
}
```

任务设计

本项目基于 M300 上 OSDK 的实时视频流功能来读取相机的实时视频流,并通过 OpenCV 实现人脸检测。在实施前,请在树莓派上使用源码编译安装 OpenCV。本项目将基于 C++完成。

任务的具体设计要求如下。

(1) 创建项目目录。

(2) 本项目基于 liveview 功能,项目目录下的 liveview 目录用于编写代码,以实现用户线程调用 liveview API 接口。

(3) 在 liveview 目录下编写代码,实现 liveview 功能。

① 初始化 liveview 模块。

② 通过 FPV 相机进行视频流传输。

③ 使用 OpenCV 进行人脸检测。

(4) 使用 cmake 编译项目。

(5) 通过 VNC 远程连接到树莓派,执行项目。

任务实施

限于篇幅,这里只给出任务实施步骤和关键代码。

1. 构建项目

连接树莓派开发板,默认路径为当前登录用户的家目录。该目录下有一个名为 pro_16_facedetect_source_template 的目录,此目录下保存了本项目的所有模板文件。

目录结构如下。

```
pi@raspberrypi:~ $ cp -r pro_16_facedetect_source_template pro_16_facedetect
```

2. 修改 dji_sdk_app_info.h

填写用户账号信息。

```
#define USER_App_NAME                          "RaspberryPi"
```

```
#define USER_App_ID                    "126413"
#define USER_App_KEY                   "0ea2eadb314d1a6e870975364e2bf4a"
#define USER_App_LICENSE               "HZ/GPtycWQSLf……."
#define USER_DEVELOPER_ACCOUNT         "xiefengran@outlo**.com"
#define USER_BAUD_RATE                 "460800"
```

3. 修改 CMakeLists.txt

设置本项目的工程名为 pro_16_facedetect，增加 CXX 选项。

```
# -------------------------设置工程名-------------------------
project(pro_16_facedetect CXX C)
# -----------------------------------------------------------
```

4. 修改主函数

在 main.cpp 中进行修改。

```
/* Includes ---------------------------------------------------*/
#include <liveview/liveview_entry.hpp>
#include "application.hpp"
#include <dji_logger.h>

/* --------在main函数中调用服务入口函数------------------*/

int main(int argc, char **argv)
{
    Application application(argc, argv);
    char inputChar;
    T_DjiOsalHandler *osalHandler = DjiPlatform_GetOsalHandler();
    T_DjiReturnCode returnCode;

start:
    /* --------调用服务入口函数--------------------*/
    DjiUser_RunCameraStreamViewSample();
    /* ----------------------------------------------*/


    osalHandler->TaskSleepMs(2000);

    goto start;
}
```

5. 修改 liveview_entry.cpp

注意：这里的文件后缀名为.cpp。

（1）添加头文件。

```
#ifdef OPEN_CV_INSTALLED
/* ---------------添加OpenCV的头文件---------------- */
#include "opencv2/opencv.hpp"
#include "opencv2/dnn.hpp"
```

```cpp
#include "opencv2/highgui/highgui.hpp"
/* ------------------------------------------------------ */
```

（2）在 DjiUser_RunCameraStreamViewSample 函数中添加以下代码。

```cpp
void DjiUser_RunCameraStreamViewSample()
{
    /* --------启动 FPV 摄像头获取实时视频流--------------------*/
    liveviewSample->StartFpvCameraStream(&DjiUser_ShowRgbImageCallback, &fpvName);
    /* ------------------------------------------------------*/

    /* --------停止 FPV 摄像头获取实时视频流--------------------*/
    liveviewSample->StopFpvCameraStream();
    /* ------------------------------------------------------*/

    delete liveviewSample;
}
```

（3）在回调函数 DjiUser_ShowRgbImageCallback 中添加以下代码。

```cpp
static void DjiUser_ShowRgbImageCallback(CameraRGBImage img, void *userData)
{
    /* ------在此处完成人脸检测 ------------------------------*/

    /* step 1 创建级联分类器------------------------------*/
    auto faceDetector = cv::CascadeClassifier(tempFileDirPath);
    /* ------------------------------------------------------*/

    /* step 2 实现人脸检测--------------------------------*/
    std::vector<Rect> faces;
    faceDetector.detectMultiScale(mat, faces, 1.1, 3, 0, Size(50, 50));
    /* ------------------------------------------------------*/

    /* step 3 将检测的人脸用方框框起来----------------------*/
    for (int i = 0; i < faces.size(); ++i) {
        cout << "index: " << i;
        cout << "  x: " << faces[i].x;
        cout << "  y: " << faces[i].y << endl;

        cv::rectangle(mat, cv::Point(faces[i].x, faces[i].y),
            cv::Point(faces[i].x + faces[i].width, faces[i].y + faces[i].height),
            Scalar(0, 0, 255), 2, 1, 0);
    }
    /* ------------------------------------------------------*/

    /* step 4 显示检测的人脸------------------------------*/
    imshow(name, mat);
```

```
    /* ----------------------------------------------------- */
    cv::waitKey(1);
#endif
}
```

6. 编译并执行项目

```
# 在pro_16_facedetect目录下，创建build目录，并进入build目录
pi@raspberrypi:~/pro_16_facedetect$ mkdir build && cd build

# 使用cmake创建项目编译环境
pi@raspberrypi:~/pro_16_facedetect/build $ cmake ..

# make 编译
pi@raspberrypi:~/pro_16_facedetect/build $ make

# 执行文件位于build/bin目录下
pi@raspberrypi:~/pro_16_facedetect/build $ cd bin/

# 执行本项目需要通过VNC连接到树莓派。在VNC的界面中使用sudo执行项目的执行文件
pi@raspberrypi:~/pro_16_facedetect/build/bin $ sudo ./pro_16_facedetect
```

执行结果如图16-1所示。

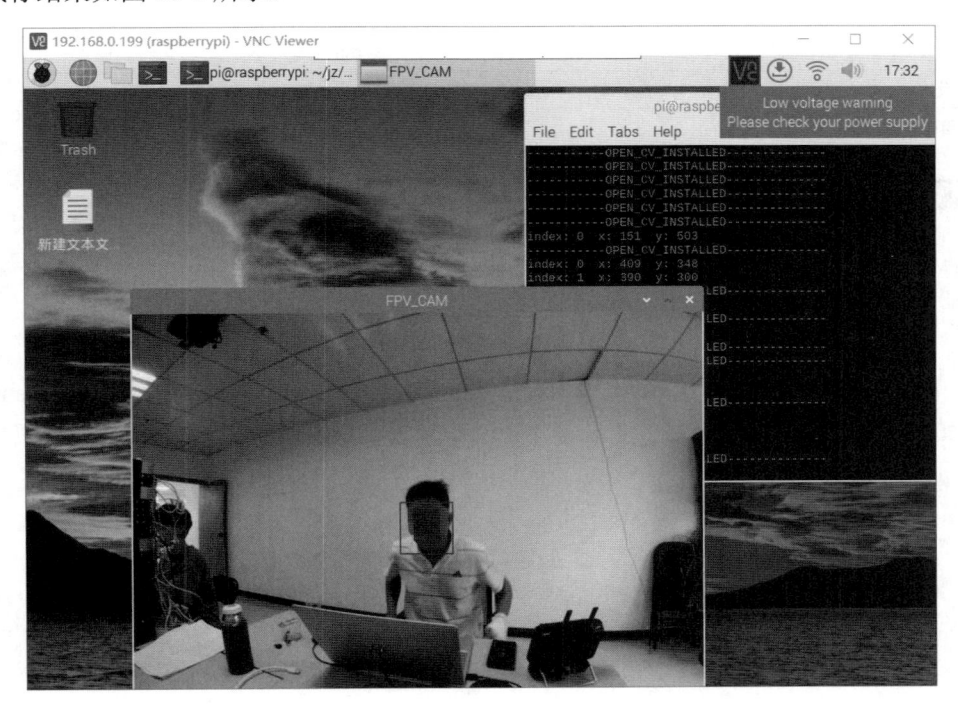

图16-1 执行结果

任务评价

任务过程评价表

任务实施人姓名_____　学号_____　时间_____

评价项目及标准		分值	小组评议	教师评议
技术能力	1. 对基本概念的熟悉程度	10		
	2. 掌握实时视频流的传输原理	10		
	3. 了解 OpenCV 的作用	10		
	4. 了解人脸检测的基本原理	10		
	5. 掌握使用 OpenCV 进行人脸检测的方法	10		
	6. 掌握测试方法	10		
执行能力	1. 出勤情况	5		
	2. 遵守纪律情况	5		
	3. 是否主动参与，有无提问记录	5		
	4. 有无职业意识	5		
社会能力	1. 能否有效沟通	5		
	2. 是否使用基本的文明礼貌用语	5		
	3. 能否与组员主动交流、积极合作	5		
	4. 能否自我学习、自我管理	5		
总分		100		
评定等级：				
评价意见		学习意见		

评定等级具体如下。

A：优，得分>90。B：好，90≥得分>80。C：一般，80≥得分>60。D：有待提高，得分≤60。

小结

本项目首先简要介绍了 OpenCV 和人脸检测的基本概念；其次介绍了使用 OpenCV 检测人脸的基本方法；最后介绍了实现基于 liveview 实时视频流进行人脸检测的代码框架。